Hotshot

To Eddy and Mary —
There's no words to say "thanks" for our friendship and your help and kindness — May we cheer, root, and moan together as we watch the Niners, Giants, and our golf game for many years.

John

To Lara and David,
who love wild places, wildlife,
and wild stories.

HOTSHOT

John Buckley

PRUETT PUBLISHING COMPANY
BOULDER, COLORADO

©1990 by John A. Buckley

All rights reserved. No part of this book may be reproduced without written permission from the publisher, with the exception of short passages for review purposes.

First Edition
1 2 3 4 5 6 7 8 9

Library of Congress Cataloging-in-Publication Data

Buckley, John, 1948–
    Hotshot / by John Buckley. — 1st ed.
      p.   cm.
    ISBN 0-87108-809-6 pbk.
    1. Wildfire fighters—United States. 2. Wildfires—United States—Prevention and control. 3. United States. Forest Service—Officials and employees. I. Title.
  SD421.3.B83   1990   90-39738
  363.37'9—dc20   CIP

Cover and book design by Jody Chapel
Cover photos courtesy of U.S. Forest Service

Contents

	Preface	*vii*
	Introduction	*ix*
1	The Clover Fire	2
2	It's a Hotshot Life	14
3	Dry Falls	20
4	A Wilderness Fireline	34
5	A Palm Springs Bash	46
6	Hotshots on Parade	54
7	Desert Wildfires	60
8	Cat Houses to Joshua Trees	70
9	Up to Idaho	80
10	The Wheeler Gorge Fire	90
11	Say Goodbye to the Wildlife	100
12	Our Greatest Fears	106
13	Spice on the Firelines	116
14	Why Firefighters Die	124
15	The Good Side of Fire	132
16	The Indian Fire	140

Preface

Turn on the evening news on any hot summer night and the odds are good that wildfire will be a top story. From Idaho to Southern California, from Alaska to Florida, fires in recent years have frequently taken center stage as media attention focused on the drama and the devastation created by windblown flames.

In a time when we can send a space probe millions of miles from earth, descend into the inky darkness of the ocean depths, or explore the secrets of the tiny atom, it both amazes and frightens many people to find that even by mobilizing every resource available, we often can't stop a raging wildfire. What is perhaps more startling is the knowledge that against many of the biggest, most dangerous conflagrations, it's often not high-tech equipment or great fleets of fire engines that finally turn the tide against the flames.

Instead, it's special crews of elite firefighters, equipped only with simple handtools and chainsaws, that aggressively cut firelines and light backfires to beat back the flame front and save homes, lives, and vast areas of natural resources.

Although great brush and forest fires throughout the West have often gained headlines as frightened townsfolk and rural residents have fled the threatening flames, the struggle to stop the incredible wildfires in the drought-stricken Yellowstone region in 1988 emphasized just how unpredictable, how overpowering wildfires can be. Backlit by firestorms crowning through the forested slopes behind them, many newsmen found it difficult to describe the terror, as well as the awe, of watching thundering waves of fire sweep ravenously across the great park and the rugged national forest lands around it.

This stunning display of one of nature's most deadly phenomena shocked many viewers by revealing how hard it is to fight fire amidst steep canyons, sheer cliffs, and almost impenetrable

Preface

forests. After watching the mostly one-sided battle, many Americans probably assumed that everyone wearing yellow Nomex-fabric fireshirts was a hero, and maybe they were. But few realized the big difference in roles played between the great majority of fire crews from state, park, and Forest Service agencies, and the special "Hotshot" handcrews.

As frightening and as strenuous as any work can be with firestorm conditions all around, most crews were only called upon to maintain previously constructed firelines or to "mop-up" burning logs and cooling embers. It was the Hotshot crews that hiked or helicoptered to the head of most fires to actually battle the hottest flames and take the greatest risks.

Unknown and unheralded outside of firefighting circles, Hotshot crews often are packed up and off to another fire long before the ashes have cooled and local communities gratefully begin to thank the remaining firefighters. But to those familiar with the Hotshots, their skills and amazing stamina reflect in many ways the best qualities of those employed in our nation's riskiest profession.

Traveling across America, Hotshot crews may face desert heat, lightning storms, snow flurries, or great windstorms, all in a single fire season. And always, as they play out the chess-like challenge of mental and physical maneuvering that wildland firefighting requires, they balance their desire to save wildlife, homes, and valuable resources with an even greater desire—the need to somehow survive amidst the flames.

—JOHN BUCKLEY

Introduction

Think back to the biggest fire you've ever seen—one where you were close enough to actually feel the heat. It might have been a giant bonfire at a high school football rally or maybe only a roaring fire in someone's fireplace after a Christmas dinner. Wherever it was, try to remember the sensations you experienced: the pungent smell of smoke in your nostrils, heat radiating intensely against your bare skin, flickering waves of orange or yellow holding your gaze as you stared deeply into the flames.

Most of us have many memories of fire. We can almost sniff that piney aroma, almost see the crackling, glowing campfire of a special summer long ago, with marshmallows slipping slowly off a crooked stick or the hissing sputter of juice dripping from foil-wrapped trout down onto red-hot coals.

This is the tame face of fire, the only side most people ever see. It's our warm friend, our efficient tool for cooking, heating, clean-up, or just entertainment. But there's another side of fire, a side that's breathtaking, frightening, and incredibly powerful.

Imagine a one-acre wildfire, with more than 40,000 square feet of brush and trees burning or threatened by the flames. The heat is tremendous, so hot that first your ears, then your face would be quickly seared. Flames leap up the tree trunks in bright, fiery fingers that flare on up through the pine needles or leaves in crackling bursts of noise. Along the ground, burning bushes pass the fire from one to another until only solitary plants remain uncharred in the midst of the burning acre.

Once you can really imagine such intense heat, take another step up and try to imagine a forest as large as a whole city block burning at once . . . then think of ten city blocks, then more. Stretch your imagination to create a mental image of a whole mountain, thousands of acres, burning with a tremendous flame front sweeping relentlessly forward through the trees and brush.

Introduction

Think of strong winds that lift burning embers high up into the smoky air, dropping them far ahead of the main fire to ignite new spot fires that grow and join together with the larger flame front.

Birds dart through the smoke in confusion, wood rats and squirrels scurry up branches or under rotted logs, and the forest itself trembles as the warm winds whip through the swaying branches. When you've finally imagined the incredible heat, the sounds and the smells, think of yourself standing in that forest ahead of the fire.

Give yourself a hard hat and gloves, canteens and a backpack, a bandana to breathe through, and safety glasses so you can open your eyes. Now join up with the rest of your crew and working together with chainsaws, shovels, and other handtools, try to stop the fire.

Welcome to the world of the Hotshots.

Literally exploding in the midst of a violent firestorm, a section of forest dies in a rush of flames.
Courtesy Bob Tribble, U.S. Forest Service

one

The Clover Fire

We stand in one long straight line in the darkness, eighteen men and two women, waiting impatiently for our plane. The initial thrill and noisy reaction to our first off-forest fire assignment of the season has slowly faded along with the last dull glow of sunset. Now the lingering heat radiating up from the asphalt runway drains away any vigor or enthusiasm still left.

Three hours ago, amid a barrage of yells, laughter, and wisecracks, we left behind wives and friends high in the heavily forested mountains north of Yosemite National Park, cramming ourselves and all of our gear into two nine-passenger crew carriers and a pickup truck. During the two-hour drive down to this hot, sprawling airfield in the flatness of California's great central valley, we lost that first adrenaline rush of excited anticipation, and now, after another long hour's wait here at the air attack base, we wonder if they've already caught the fire. Maybe all we have to look forward to is a slow drive back home.

Up at the head of the line, our superintendent, affectionately known as the "Big Guy," suddenly laughs a long, loud abrasive laugh that echoes back from the huge tanks of fire retardant that loom behind us against the perimeter fence. Boss of our Hotshot crew, though not much older than many of the other firefighters, the Big Guy elbows one of our squad bosses in the ribs as he reminds him of some ribald jest.

"I guess he's not worried yet," I murmur softly to Larry, making sure my voice doesn't carry too far. "He must think we're still going."

Although I'm an experienced firefighter, this is my first season on the Hotshots and I still feel my rookie status. In the midst of fifteen veterans back from previous seasons of arduous but glamorous firefighting, those of us who are new have already tasted too often the bitter ridicule of foolish questions like "Why is the plane late?"

Larry nods his curly black Afro haircut affirmatively, then stoops once again to quickly sort through his assembled gear, making sure for the last time that nothing vital has been left behind.

"My ballcap . . . where's my ballcap?' he mutters, pulling out a growing pile of gloves, sunglasses, hard hat, and handkerchiefs before Nell's amused laugh cuts him short.

"Try looking behind your big pack," she points, calm and relaxed in the confidence she wears from two years of experience with this crew. "Don't worry," she soothes, "if you've got all that fire gear the only other thing you really need is your boots, and you're wearing those, so relax."

She smiles amusedly as she watches Larry still searching nervously through his stuff, then turns to me with a wry explanation for the plane's delay. "Remember," she teases, "that government agencies do everything by low-bid contract. Even the planes we fly always come from the companies with the lowest bids."

"Once we waited here all night . . . the plane finally showed up the next morning." Then she adds, on a more serious note, "But they're almost always an hour or two late. You watch, it won't be long now."

Her jest about the "low-bidder" plane doesn't bother me. I've flown off-forest a couple times before this as part of regular Forest Service firefighting crews, but Larry groans even more nervously. "I can hardly wait," he sighs as he stuffs his gear tightly back into his pack for the last time.

He doesn't wait long, for minutes later someone spots another plane descending through the bright stars that cover the sky on this moonless night. "That looks like it," Nell comments, and sure enough, the plane taxis on past the main airport and drones loudly on down the runway toward where we wait expectantly in the darkness.

From a trailer-office parked between the retardant tanks and the large warehouse, the sleepy base manager hurries out to the Big Guy to pass on last minute information. Then the roaring Convair plane slowly pulls even, swings directly toward us, then turns to point back the way it came, stopping only a hundred feet from our eager line.

More or less like a single unit (a little less this first time for the season), we move forward to the plane, passing all our big packs and daypacks forward in a stylish "chaining" maneuver designed not only to keep us from crowding the tight quarters under the storage compartment of the plane, but also to ensure our appearance as professional. Then it's up the stairs into the airplane, filing neatly one by one into the available seats—those left in the rear half of the plane by the first crew.

The interior of the plane is dark, but in the intermittent reflection of the flashing lights from the plane's exterior, I can catch glimpses of the other crew.

A single stewardess briefs us with the standard routine of seatbelt precautions and emergency exit explanation, then we're quickly taxiing down the runway. The plane is old, old enough to make me wonder for a brief moment about the possibility of crashing. We taxi for a long ways, bouncing stiffly over each bump to our designated take-off position.

Amidst a lull in many low conversations, the pilot jams back the throttle, the plane leaps forward, and we're speeding back down the runway . . . still touching . . . still touching . . . then off and rising up in a long steep turn that takes us south over the thousands of tiny lights that seem to fill every inch of the broad valley beneath us.

Even more than the waiting, the lighting of the plane and the late hour begins to take its toll. I can feel the need to sleep tugging my eyelids shut, although the ride is only to Bakersfield and a few of the crew are talking just loud enough to keep everyone else awake.

At last the jokes and bragging quiet, and both crews seem poised to sink back into their seats to sleep. But already we've begun our gradual descent and only minutes later the stewardess is checking our seats for the landing position.

Nowhere on the horizon can I spot the glow of a rampaging wildfire.

From the airport we take a big commercial bus up into the mountains. Everyone's tired and more than a few are grouchy. It's after midnight and all of us have already put in a hard day's work cutting brush and small trees on fuel breaks before we got our dispatch. Even in the cramped quarters of the bus we manage a catnap here and there, but the constant turns and shifting as the driver winds up the narrow mountain roads makes real sleep impossible.

By three in the morning, we're "there," at firecamp, wherever that is, for except for a couple of trailers, some lights, and a generator, there's nothing here.

We pile out, unscramble all our gear and tools for the last time for the night, then raggedly march single-file like a band of overloaded African porters over to a sandy area in the midst of creosote bushes and sagebrush.

"We must be on the east side of the Sierras," Denny, my squad boss, points out as we quickly spread our sleeping bags on the rock-strewn ground. "Everybody get to sleep as quick as you can—we'll be getting up at four-thirty."

I glance down at my watch, knowing I shouldn't even as I do it. It's three-thirty now. That means maybe an hour of sleep if I'm lucky. For the first of many times, I question why I chose to be a Hotshot.

"Get up!" A foot nudges me roughly in the shoulder. It's dark and chilly. The boots I seem to have just taken off feel ice-cold as I struggle to get back into them. Then lined out, as if on parade, we march over to the newly arrived chow-wagon.

Breakfast isn't ready yet, but we wait dazedly, blinking our sleepy eyes as more buses steadily whine into firecamp, spilling out forest crews from nearby national forests and California Division of Forestry inmate crews from conservation prison camps in the area. We're first in line, with the Lassen Hotshot crew that came in with us last night right behind us, and by five-thirty we've eaten and regrouped at our sleeping area. In a last quick briefing, Denny outlines some additional directions to our squad, the tool squad, while the saw teams check over their equipment and fine-tune their chainsaws.

Minutes later, Paul, our foreman, signals to us to get on the bus, where the Big Guy joins us with the Lassen crew. Fire-ready, we drive a mile or so back down the dirt road to this fire's designated heli-base, a flat open area with good access in and out for the steady stream of helicopter flights soon to begin.

Then we wait, struggling to stay alert in the glaring morning sun, waiting for the final decision on who flies where and when. Already it's warmed up considerably, and by the time the first observers roar off in a clatter of whining blades, anyplace in the direct sun is uncomfortably hot.

"We fly after Lassen" comes the message, and so we retreat back into the partial shade of some clumps of scraggly bushes to wait for our turn. A few of us try futilely to sleep, but even as tired as we are, the incredibly high noise levels of helicopters landing and taking off only 200 feet away keeps us awake. Then, too, the unknown lies just ahead. It's impossible not to speculate on what the fire is doing.

Firecamp gossip, always taken as gospel until we hear differently, sets the size of the fire at 400 to 500 acres. If true, and if the fire's edge fingers in and out like that of most fires driven by strong, erratic winds, the outside edge of this fire is already more than four miles around. We've also heard that so far the fire is almost completely uncontained, with only a single crew of local forestry personnel working somewhere along the bottom section of the fire.

From the heli-spot we're blocked from actually getting a clear view of the fire by a large ridge and the shoulder of the steep mountain it's burning on, but drifting clouds of gray-white smoke billow up steadily from somewhere along its flank.

"It's probably just creeping around," seems to be the common consensus, although calm winds down here don't mean calm winds up on the mountain.

The last of the Lassen crew boards an S-212 copter and we struggle back to our feet, hoisting our tools, daypacks, and web gear (which holds our canteens) out of the shade and stacking them in neat piles, ready for the next helicopter flight. Yet even as the first third of our crew files forward to wait beside the helitac crewman who'll direct their loading, a new order to shut down the helicopters blares over the loudspeaker mounted on one of the trucks.

Like yo-yos on a string, we grab our gear and trudge back to the diminishing shade of our bushes, unsure of how long this shutdown will last. Through the fault of no one in particular, we've gotten less than an hour's sleep since waking up a day ago, and despite our rush to breakfast and then over to this heli-spot, another three hours have slowly passed.

"Hurry up and wait!" moans Denny. "The story of a Hotshot crew . . . it comes with the job."

I nod, then shrug my shoulders. What else can we expect? High above the smoke I see the reason for our delay, as sunlight reflects off a big DC-6 drop plane as it circles the fire in preparation for its retardant drop.

A second drop plane, smaller and capable of holding much less fire-retardant liquid, flies toward the mountain from the west. Then we can spot the lead plane as it swoops out from behind the mountain and dives, wiggling its wings as it drops down in the flight path the lead plane pilot will expect the other two drop planes to follow.

The DC-6 circles wide behind the mountain, reappearing moments later to dive down out of our sight behind the ridge. Minutes later the smaller plane copies his route before both planes swing back toward the west, leaving only the lead plane to circle slowly, high over the mountain.

Even as the planes disappear, one of the big Sikorsky helicopters roars to life, its hammering staccato beat winding up in intensity until its whine is deafening. I see heads shaking again as we hear them shout out the names on the next flight. Not many of us would choose the big, clumsy-in-appearance Sikorsky over the other sleeker, more maneuverable helicopters at this high elevation. But, of course, no one's asking us.

The first part of the crew is already hurrying forward, carrying their gear in their hands, knees bent and squatting a little as they near the helicopter. We've all flown before and been well drilled on the dangers and requirements of helicopter flights, but the roaring whine of the spinning blades ripping through the air inches above your head always seems to be an especially potent reminder to duck down or keep to a crouch. Stories of loppedoff heads, although never personally verified, always insure plenty of respect for the nearly invisible blur of the rotors.

I grab my daypack, slip into my heavy web gear, and move in line behind Larry. The scream of the Sikorsky's blade moves

The Clover Fire

another few pitches higher up the scale and then the ship rises up, blasting those of us still waiting with a stinging cloud of driven sand and tiny windblown bits of debris.

The second helicopter, the S-212, lands, returning from somewhere to the east, and the second part of our crew boards. Those of us remaining bow our heads toward the ship, protecting ourselves with our hard hats as the copter lifts off, pelting us again with stinging debris. For a minute or two, the sudden silence dominates, then we're distracted by more buses bringing crews down the dusty road. Even as the first bus begins to spew forth two forest hand crews, the Sikorsky rumbles back into sight, gradually descending to vibrate and shudder before us expectantly. I tug my chinstrap down under my chin, readjust my sunglasses tight to my cheeks, and move forward in a low crouch behind the heli-tac attendant.

He points to a seat beside the door as I step up and in, so I snap the seat belt open, settle in, re-snap the seat belt, pull my knees back, and grip my gear tightly in both hands. As soon as all seat belts are clearly fastened, he backs off, leaving the big sliding door open as they do on many shuttle flights, and signals to the pilot with a thumb's up gesture.

We lurch upward and forward in less than fancy fashion, climbing quickly up and over the scattered sage desert toward hills dotted with pinyon pines and junipers. Amazingly enough, I can see larger pine and fir forests ahead as we climb steeply up the main mountain. Below, the blackened evidence of charred hillsides and still-smouldering trees shows yesterday's damage.

We swing up a valley, over a ridge, and across another valley and there, below us, is a steep river canyon. The fire's burned in many directions without leaving a clear pattern. The thick smoke hanging in dense clouds makes it hard to tell if there's any one main head to the fire or only a lot of minor problems.

The mountain looms up right beside us as the copter pilot angles our route toward the very top. Lush green meadows and heavy forest cover all the higher part of the mountain, and even though I've seen many beautiful mountains in my travels, I'm especially impressed by the steep, rugged splendor of this panoramic scene.

Just as my concentration shifts from the fire behind us to

the beautiful scenery of forested meadows, lakes, and peaks on beyond our mountain, the copter shudders as the pitch changes and we descend toward a long, narrow meadow tucked down in a blanket of thick, old-growth forest stretching across this wilderness height. Yellow dots below gradually turn into the rest of our crew, waiting just off the meadow's edge in the cool shade. Then the ground seems to rush right up at the open door of the ship and we land softly in the still-moist and wildflower-strewn meadow.

Once our crew is intact and the ship has gone, we grab a quick bite to eat before leaving. The Big Guy tells us the fire is reportedly closer to 1,000 acres already, has two distinct flame fronts, and, right now at least, isn't doing much.

We gear up, untaping tool edges and revving up our saws, then we line out through the forest in our firefighting alignment. Ten minutes later, we drop out of the vacation-like paradise of the mountaintop into a choking fog of thick, stinging smoke. Great boulders and steep dropoffs hinder our movement as we work our way single-file down through the fairly thick forest to pick up the fire's edge somewhere down the slope.

A loose rock here, watching out for slopes of slippery pine needles there, we drop still farther down off the top of the sharp ridge we've followed down this far. A steady breeze is keeping the thick smoke pushing into us, but we're still not right at the fire's edge.

Finally we drop down and down and suddenly, the whole forest on one side of us is burning. Great rotting logs flare up with tremendous heat and a ragged fire edge of spreading flames moves down below us through the forest litter of pine needles, ground covers, and branches.

Three saws start with a high whine, one following the other as they cut a path through the brush and smaller trees twenty feet or so off the active fire's burning edge. Then comes the rest of our crew, wielding sharp-edged shovels interspersed with axe-like Pulaskis and scraping tools to cut through the ground covers and remove all burnable litter in a four-foot-wide fireline stripped down to mineral soil.

We anchor the fireline to a large rock outcropping, then work our way down along the fire's edge, chopping and flinging dirt at the flames. The radiant heat of this fire scorches my face, the only exposed skin on my body, so I quickly wet my handkerchief and tie it over my mouth and nose. Then chop, scrape, chop, scrape, over and over, again and again, down and down.

Firefighting with handtools is much like digging a shallow ditch next to a bonfire, working at a fast pace and sucking far too much smoke. Our saw teams have pulled out away from the rest of us already, for there's not a great amount of brush to cut out through this section of forest. Far down below I'm sure I can hear the roar of the river.

I yell something about listening for the river to Larry, then I slide on down a particularly steep rocky slope to pick up the line construction again. Denny holds his handi-talki radio in close to his ear and turns a little, stopping at the point where reception is clearest. He shouts something from up above me at the end of the line, and quickly I hear the order "Stop work! Bump down the line!" passed down through the crew.

Immediately I shout the message on down to Larry and those ahead of him, then hurry to get out of the way of those sliding recklessly down at me from above. We all join together 200 feet below, where the Big Guy stands tensely, ordering us to hurry into the heat of the burn.

There must be a blowup far below us, for why else would we force ourselves into the stifling heat but comparative safety of this partially burned forest with giant trees still on fire all around us and the burning ground so hot it scorches our feet right through our thick-soled boots? I can hear the roar of the river much closer now, and wonder why we don't just drop on down to it.

Then the roar increases and I realize my mistake. That's not the sound of a river. It's the roar of a massive firestorm rushing up through the forest below us. Already strong winds sweep through the pines with great gusts, and I can see the forest below starting to glow with an eerie red light reflecting through the dense smoke.

We huddle in a small rocky clearing surrounded by burning snags and smoking ground covers. Minute by minute the firestorm grows both in crescendo and velocity of the wind.

We're a hundred yards or so from the unburned edge of the

fire, but even here a re-burn could sweep through the remaining trees and unburned brush with a deadly wave of flames that could fry us in a minute. The ground is so hot that we hop from foot to foot in futile efforts to cool our feet. The order comes down, "Don't wet your boots!" and I remember stories of burned feet from the steam produced by firefighters who thought the water might cool their sweltering feet. Besides, every drop in our canteens is vital, as we sip constantly to soothe parched throats.

I shift from one foot to the other, searching constantly for a place to stand where the ashes' heat won't penetrate so painfully. A sudden rush of hot air blasts into our faces and I can barely make out the forest about us in the darkening smoke.

Then the full force of the firestorm hits the unburned slope we just labored so hard to save with our fireline. I try to decide if the thunderous rush and roar of the flame front sounds more like a speeding freight train or the thunder of ocean surf, but can't quite match it with a good comparison.

I pull my handkerchief even higher on my face, leaving just a slit for my eyes to peek out at this incredible phenomenon. The firestorm pulsates with such thunderous bursts of explosive "runs" through the preheated forest that sudden gusts of heat blast across the slope to cover us with whipping clouds of ashes and embers. Wave after wave of brown-orange fire shoots up into the few trees visible through the smoke as more embers and whipping pine needles cascade in every direction. No one says much. We feel no immediate threat to our lives, but the very thought of anyone surviving out where we'd been minutes before seems ridiculous.

The main thrust of the firestorm peaks after fifteen minutes or so, slowly sinking from the thunderous roar down to a still-awesome crashing and crackling as crumbling trees disintegrate and branches fall in fiery arcs. Once we can hear and talk over the noise of the fire, a steady stream of complaints and grouching echoes from one cluster of hot-stepping men to another.

For almost another hour we struggle to breathe and not bake in the dense choking smoke and sweltering heat of our chosen sanctuary, fearful lest a second wave of fire should sweep up through our still partially unburned section and endanger us again.

At last, just when it seems that all oxygen has disappeared from our bodies and our lungs will never withstand another five minutes of this gagging smoke, the Big Guy lines us out and we carefully retreat back up through the burn. Not surprisingly, I find that the short time it took us to climb down the mountain is considerably lengthened as we attempt to weave our way back up through the burning logs and up across near vertical rocky bluffs.

An exhausting hour later we reach the top to find the fire has swept up over the sharp ridge and slopped over into the fairly flat forest on top. We take a short rest to catch what we can of our breath, then it's back to firefighting again as we pick up the now slow-moving fire's edge and gradually work our way back toward the string of meadows where we landed.

Soon we're joined by an equally fortunate Lassen Hotshot crew who weathered the same fire run farther down the mountain. Between our two crews, we quickly build a quarter mile of secure, well-constructed fireline, and the last fading rays of golden alpenglow find our crews tying in the fireline to the unburnable border of wet, narrow meadows cutting across the very heart of the mountaintop.

Totally exhausted and famished from our exertions, we slowly stumble up through the meadows to our original landing spot where a semblance of a spike camp has been hastily flown in to feed and supply us for tomorrow. We line out to eat, scooping huge portions of still-warm potatoes, steaks, and vegetables onto paper plates, then there's still all the tools to sharpen, saws to clean, and equipment to check before we can sleep.

Just before midnight, all lies in readiness for tomorrow. I eagerly grab a paper sleeping bag from the great pile of boxes flown in by the copters. Regardless of prodding rocks, possible snakes, or recollections of this day's potential dangers, I slip down inside, close my eyes, and instantly I'm asleep at last.

Moving out ahead of the fire, the toolers follow where the saw teams have already cut, removing what's left of flammable fuels along the fireline.
Author's photo

two
It's a Hotshot Life

Sunrise the next morning finds our crew trudging sleepily, but well fed, up a long ridge that connects our blackened mountain with its neighbor a mile or so away. Along with other crews, we're attempting to outflank the fire by constructing firelines well ahead of the fire's present slow-moving advance. Then, when conditions are right, we hope to backfire off our fireline, burning all the forest between our line and the main wildfire. With a little luck and the sacrifice of a considerable amount of forest, we may actually be able to catch this fire before it spreads to the next mountain.

Water was scarce when we left camp this morning, so the sudden appearance of a sling-load-carrying helicopter brings shouts of relief up and down our line. We can work with minimal food and equipment, but handcrews like ours can't perform long in the desiccating heat of summer fire weather without plenty of water.

The copter circles once, hovers so that the sling load almost brushes the ground, then releases the sling and soars back off toward the spike camp. We converge quickly on the drop site, scuffling good-naturedly over the four gallons of milk that some kindhearted service chief back at firecamp sent up with the plastic "cubi-tainers" of boxed water.

Even as we gulp the milk, the Big Guy and the Lassen superintendent are yelling at both crews to "water up" and get going. We hurry the task, spilling precious water all over the rocky slope as we try to pour from the three-gallon containers into the four to six small quart canteens we wear on our web belts.

The smoke from the fire is still drifting downslope and away from us, so the bright sunlight and far-reaching views from this ridge give us a lift as we regroup and climb across to our starting point. The Lassen crew drops off at a rocky knob to begin their section of line construction while we trudge far on down the ridge to another point where thick forest blankets all but a thin strip on the very crest of the ridge.

The saws roar to life, we settle into our established tool order, and we're off, cutting and scraping for hour after hour in the steadily rising heat. As noon approaches, we're close to tying our line in to fireline already built by a regular forest service crew flown in by helicopter early that morning. Radios blare fire-talk back and forth from firecamp to spike camp to crew supervisors and finally the "OK" comes down to start the backfire.

As we spread back up our line (each one of us covering a hundred yards or so of line to watch for spot fires or creeping escapes), we can see great dark-gray clouds of smoke spiraling up from the backfire the Lassen crew is setting. Our burners work their way along our line, lighting not only the flammable ground covers right along our fireline's edge, but also enough bushes and trees inside to build up a heat source that will draw the fire away from our narrow fireline. At first the burning goes perfectly, with all the flames backing away from our line just right and consuming enough small trees and bushes to ensure that no re-burn will cross our line. Then the winds get gusty, probably from the conditions created by air rushing in toward the tremendous column of smoke and flame building up from all the many backfires now burning.

First a few embers rain down across the line, only to be quickly extinguished by scurrying Hotshots. Then embers cascade down everywhere and it seems like we'll surely lose everything we worked to build. Just as a couple of small spot fires flare up in between firefighters the first drop plane roars by close overhead.

Minutes later he's circled again and we all face toward his steadily growing shape. He swings in over the forested knob, tilts one wing just a little to one side, then back, and "poof" . . . a long red cloud of pungent ammonia-like liquid rains down in a sudden shower that drenches spot fires and firefighters alike in slimy red fire retardant. I can feel the sticky red goo on my neck as well as all down my right shoulder and side, but now is not the time to worry about appearances. Quickly, amidst curses and laughter, our crew moves to take advantage of the drop, jumping onto all the smouldering spot fires and burning embers that had threatened to escape moments earlier.

We finally complete the successful burning of our section of line and get the welcome word to spread out and hold our area. For the first time since before dawn, we take a good break, digging into our bag lunches and whatever tidbits that Hotshots stockpile in their packs for hungry moments. Nell's crunching on trail munch, a mixture of nuts and dried fruit, while Larry's savoring a mouthful of candy. Somehow, it seems there's always a can of smoked baby clams going around the crew, not exactly the tastiest flavor to accompany dried apricots or chocolate kisses. Soon the flavors all blend together anyhow.

Our rest is suddenly cut short. "Line out! Let's go!" bellow the Big Guy and Paul simultaneously. "They've lost the fire farther down the line!" Stuffing sandwiches in our mouths as we struggle back into our packs, we march double-time for ten minutes on down along the next crew's newly built fireline toward the problem spot.

Sure enough, the fire's making a run out into a brushy flat saddle where strong winds are pushing flames almost horizontally along the ground. The forest crew we've followed down looks busy with a small flashy spot fire nearby, so we haughtily pass them by and speed toward the much bigger escaping slopover. When we're still hundreds of yards away I can hear Hotshots ahead of me grumbling with scorn and ridicule about "worthless blue-card crews" and as we reach the fire area I quickly understand why.

Almost forty firefighters are standing helplessly by as the fire burns its way farther out into the sage and brushfield, spreading out of control. Four or five of them are futilely flinging dirt with shovels way too far away from the head of the fire to be doing the least bit of good, and the rest seem totally confused.

Without even a word our twenty-man crew marches right through their midst and out into the path of the flames. Running in and out of the gagging smoke and roasting heat, we attack the hot spots with a savage disdain. In, then out, then in, holding our breath until our lungs must burst from the strain. One bush in front of me flares up, then all those nearby also catch, shooting flames right across my hips and shoulders and driving me back, gasping for fresh air. But Denny and Nell and Paul join me in another attack, coordinating their efforts with well-flung dirt and synchronized timing to knock down the hottest of the flames and halt this most active flank of the fire.

Fifteen minutes after we arrive, we've not only cut off the spreading flames, but widened our new fireline enough that the "blue-card" crews suddenly get motivated and jump in to help. Paul and Denny both verbally rip the newly brave firefighters with some nasty sarcasm before we line out again to leave the mop-up to them.

Then it's back up the hill to hold our own line and finally, for real, to sit down for a few minutes and reflect a little on the past two days. Denny moves up the line with the welcome message that the fire now has a line all the way around it—the hot spot we've just caught was the only escape in the whole backfire and burnout operation.

I can't help asking Denny to speculate on how many more days we'll be here. The longer we stay, the more overtime we'll make, and we can all use the money.

Denny laughs and gives me his "impatient teacher" grin. "Hey! If you haven't heard it before, remember. . . . We're the Hotshots! We may get a mop-up shift or two, but mostly they want us free, ready to go someplace else and do it all over again."

"Look at this afternoon," Denny goes on. "We had to run all the way down the hill to save the fire for two crews that get paid the same money we do. Why? Cause we're Hotshots. It's not just that we've got more experience and more training . . . a lot of it is just a matter of pride. It may not pay the bills, but no matter where we go, we know we're the best." For Denny, that's quite a long speech.

He winks as he stands up to continue along the line. "Besides, think of the glamour. You're a full-fledged Hotshot now."

I turn toward Larry, who's basking in the glow of Denny's tribute to our crew. Larry's face is covered with black charcoal and gray ashes, especially around his mouth. His teeth are grimy, his eyes are beet red from the smoke and lack of sleep, and his clothing and hard hat are heavily splattered with great gobs of dried pink retardant.

Without even the slightest doubt, I know that I look the

same or perhaps even worse than he does, and appearance is nothing compared to how I feel. My lungs ache from the smoke and despite our intense physical training program, I'm sore all over. My nose is scorched from the flames and sun, but it's still working well enough to let me know that all of us could use a shower. I can't help but laugh out loud, then laugh even more at Larry's totally puzzled expression.

"Yeah, the glamour of being a Hotshot. It's unique, that's for sure. I just hope a little of it washes off."

A Southern California "gobbler" burns in terrain where only Hotshot crews can go safely.
Courtesy Frank Podesta

three

Dry Falls

Our shuddering airplane drops suddenly as it hits an invisible air pocket and all four Hotshot crews clutch their seat arms instinctively. This flight is quickly becoming an ordeal instead of a pleasant sunny flight down the state to our latest fire assignment.

Even before we left Stockton's airport, the scorching heat and desiccating winds had prompted many predictions of a really big fire. But not until we approached the Tehachipi Mountains did the plane begin to really bounce as strong winds buffeted us erratically. Now I can actually see the wings flex as we drop again, then climb back up on an airborne rollercoaster ride. The oil streaming slowly but steadily back along the shiny aluminum casing of the nearest engine bothers me, and I grimly point it out to Dan, who's seated beside me. He shakes his shaggy head in both disgust and apprehension. "It feels like this thing's shaking apart. Damn, I hate flying in winds like this!"

A murmur of discovery flows back through the plane and all eyes turn toward the southwest. There, on a mountain range rising up sharply from the edge of the flat desert, a great brown cloud reaches high into the sky. The winds have bent the column over so far that even from thousands of feet up we can eventually see pinpoints of orange flames along the fire's spreading flame fronts.

"Looks like we're back in South Zone for real!" laughs Nell from the seat behind us. Dan just shakes his head again as the plane shudders quickly and we feel our stomachs jerk up toward our chins. "I hate South Zone," he grumbles.

Fighting wildfires in South Zone typifies the greatest danger, the most tedium and frustrations, and by far, the zaniest, most unpredictable aspects of wildland firefighting. South Zone, as the southern half of California is referred to in Forest Service firefighting management, takes in all the open oak parklands, high mountain forests, and vast, dry scrublands from Monterey to San Diego. It includes the southern Sierra Nevada, with peaks over 14,000 feet, and the sizzling-hot desert that stretches south to the Mexican border.

Within these boundaries lie thick blankets of coastal brushfields that stand higher than your head and grow so densely that even crawling, you can't penetrate more than a short distance. Oily native bush types, highly flammable grasses, pine forests covered with volatile sap, and a wide variety of introduced, hot-burning ornamental trees like eucalyptus fill the foothills and mountains. Then farther east, dry, brittle desert plants, yucca, and cactus cover the drier desert zones.

The string that ties all these different vegetation zones together into one incredible challenge for fire agencies is the climate. Fire season begins with the last spring rains in April or May and lasts as late in the fall as Thanksgiving or even into December. During this four-to-seven-month-long stretch, almost no rain falls, the hot sun warms and dries the grasses, bushes, and trees a little more every day, and strong gusty winds blow steadily in local areas. Occasionally, incredible winds called Santa Anas whip across the whole region with gusts above 100 miles per hour.

All this heat, wind, and volatile fuel types, mixed in just the right portions, produce enough of a great fire hazard without the population element. But combine them with a regional population numbering more than fifteen million, and true chaos can, and does, occur.

Holiday weekends or even just beautiful midsummer weekdays may bring millions of city folk scrambling out over every available bit of national forest, county parkland, private ranches, or any other open land. Dune buggies, trail bikes, four-wheel drives, recreational vehicles, and travel trailers wind incessantly up the highways and side roads into the steep canyons and foothills that surround the Los Angeles basin.

Crowd tens of thousands of campers into the canyon bottoms along any stretch of flowing water and eventually a wildfire will start climbing tinder-dry hillsides, heading up and over the highest ridgetops. Such accidental fires pose enough of a danger without the addition of one of Southern California's best-known "crazies," the arsonist.

Arsonists apparently hibernate for most of the year, only to emerge when brown dry grasses and parched foliage wither in week-long heat waves. Suddenly, arson fires pop up at random, often hundreds of miles apart. Then, as media coverage details their cause, "copycat" arsons imitate the first wave of fires, until cool weather or a lull in the winds finally sends the arsonists back into hibernation.

A terrible history of catastrophic wildfires in this highly populated area has created the largest and best-equipped wildland firefighting agencies in the world. Facing thousands of potential conflagrations every fire season, city, county, state, and federal fire agencies coordinate with ultra-modern computer-based dispatching, a small armada of precision-accurate retardant drop planes, and highly trained firefighting personnel.

But even with such a force poised to spring into action, it's impossible to stop every fire before it spreads out of control. Once fires burn beyond the reach of roads and on up into the sheer cliffs and eroding hillsides of the rugged mountains where bulldozers can't go, then only helicopters, drop planes, and hand crews are left to somehow stop the flames. And the top-rated hand crews are the elite Hotshot crews of the U.S. Forest Service.

So once again we're flying over the Tehachipi Mountains into South Zone, peering out the plane windows at the dark, wind-whipped cloud of smoke, trying to get a good idea of what we'll soon be facing on the ground. The plane heaves to the left so sharply that an audible groan escapes a few unprepared passengers. Immediately snickers and a chorus of imitating moans and groans sweep over the eighty firefighters until one clown covers his head with his hands and shouts, "We're all going to die!" in mock despair.

Laughter and more jests continue even through the midst of our lurching approach to Palm Springs, slowing a little as we drop sickeningly in descending jerks toward the airport below. Despite the roughness of our flight, the actual landing is anticlimactic, and at long last the comforting sensation of contact with solid ground wipes away our forced, tense smiles.

Even as the plane taxis into position we can see the out-of-control fire blazing farther up on the mountain. The steep slopes loom high above the town to the northwest and wrap around it to the west and the southwest, rising up into ridge after ridge of nearly vertical arid wilderness.

We file gratefully down the ramp and unload our gear, combining crews to "chain" all our equipment toward the runway's edge. A television camera sweeps back and forth and news-hungry reporters prod for stories. Nell and Ann, always standouts because few Hotshot crews have any women, pull their ballcaps lower over their faces and busy themselves with assembling their gear.

"What's the matter?" I tease, "Afraid word will get out you're vacationing here in Palm Springs for the week?"

Burnt out on exclusive coverage of "how women can actually fight fire," both Nell and Ann glare at me good-naturedly. Then we're waiting in the wind and sun, standing impatiently but typically for an hour before buses come to drive us to a local high school a few minutes away.

Unexpectedly, we're told we'll be bedding down for the night instead of heading straight to the fire. An air of confusion and uncertainty dominates the hastily assembled but fast-growing firecamp based between school buildings and the various athletic fields. Most of us feel a let-down after psyching ourselves up for a head-on confrontation with that gobbler of a fire. We wait around, half expecting to hear orders change and find ourselves heading to the fire after all. Paul brings us back the few rumors he can garner from the "plans" section.

Dusk plus the tremendous winds apparently prevent us from flying up on the mountain by helicopter, and except for a tramway that's already threatened by the fire, there's no other access to the upper slopes. We pass the night fraternizing with all the other Hotshot crews that managed an assignment to this fire. This is the only big blaze currently burning in the state, so more than twelve Hotshot crews are already here and most of the rest will be arriving soon.

Hours later, in what still seems to be the middle of the night, it's "Get up!" and we're fumbling in the sleepy darkness to gear up and line out to eat. After a quick bite of food, we find ourselves, along with three other Hotshot crews, squeezing tools, gear, and crews into school buses. Then in the glaring bright light of a summer morning in Palm Springs, we're driving through the town, past what some informed, impromptu tour guide describes as Bob Hope's hillside home, and winding slowly south and up into the rugged mountains.

Eventually we pop out of our uncomfortable confinement to reassemble on the edge of sunny, attractive meadows and scattered forest at the base of the back side of the same mountain range we've just circled. Then it's "Hotshots on Parade," eighty firefighters and a few strictly supervisory personnel marching single file past a solitary photographer toward the white, drifting smoke just visible over the mountain crest.

"This is crazy," gripes Larry, turning to look back at me as we start up a steep, narrow trail through oaks and pines. "We're at least ten miles away from where we saw the fire last night; it can't possibly have burned this far."

Yet it obviously has, for even as we labor for a long hour's climb up the switchbacks and duck through the overgrown brush that obscures the trail, we can see the smoke rising closer and thicker beyond the ridge. Finally, near the top, we leave the trail and struggle to climb straight up until we crowd shoulder to shoulder onto the narrow ridgeline that's choked with fifteen- to twenty-foot-tall brush and oaks. Behind us, the brushfields we've just climbed up through are nearly impenetrable except for the single, overgrown trail we've managed to follow most of the way before leaving it where it angled back toward a saddle to the south of us.

The overpowering smell of sweating bodies and the heavy panting of breathless climbers permeates the ridgetop. Our view down the eastern side of the mountain is limited to a few glimpses through small openings in the towering brush and thickets of oaks. From behind us, late morning winds blow steadily up across the ridge and down through the sea of waving vegetation, pushing the smoke from the fire down and away from us. No one knows for sure exactly how far away the fire's edge is smouldering, temporarily slowed by the downslope breezes, but a sudden wind shift and we could find ourselves eating fire with no place to escape. I don't like standing here, even to catch our breath. We'll be safer once we start cutting a wide fireline.

Shiny "overhead," or supervisory, hard hats all cluster at one side of our crowded clearing. Fingers point, heads nod, and soon the four Hotshot superintendents work their way over to learn the final decision for our deployment.

Paul and the Big Guy start moving our crew off through the bushes to another narrow opening right at the low point of the ridgeline. Three or four of our crew linger behind for a few minutes, savoring their rest after the long, steep climb. Immediately, the Big Guy pushes back past us to chew them out mercilessly. Sulking, they catch up with the rest of us and the Big Guy gestures for all of us to crowd in close. We wait reluctantly while he pulls a large wad of tobacco out of his mouth and casually spits the remnants at his feet. Then he begins to lay out in detail exactly what he expects from us. He tells us we've been too slow, complained too much, and talked too much around the overhead. What's more, he's already thinking of getting rid of some of us — exactly who he doesn't say. We've had it easy so far and he expects us to jump to and look sharp, or else!

After a few more complaints, he spells out our assignment. We're to cut a swath fifteen feet wide with a four-foot clean scrape up along the top of the ridge toward a gray rock outcropping a mile or so ahead. Another crew will come behind us and widen it out to twenty or twenty-five feet, hopefully wide enough to contain the slumbering fire once it awakens again.

We line out slowly, knowing it'll take awhile for the saws to open up the canopy enough for the rest of us to follow behind. Instead of just working in teams of one sawyer and a puller (who throws the cut brush or tree limbs away from the fireline's edge), we send idle toolers forward to help out. The going is so thick at first that the plans quickly change and two saw teams from the other crew also move past us to leapfrog with our sawyers. The ridge is narrow and rocky despite its dense cover of brush and oaks. Great boulders and broken, slanting rock lie interspersed with foot-deep leaf litter and duff. Gradually the five saws and their entourage of pullers move away as the rest of us dig and scrape in frustration, tediously attempting to remove all possible fuels from the rocky, duff-covered center of the canopy cut.

The higher we climb the ridge the better we can see, and the view is quite a contrast to my expectations. Instead of the desert-type sage and mesquite that fueled the fire as it roared up the mountain from the edge of Palm Springs or even the oak thicket that now surrounds us, these upper slopes reveal totally different zones of vegetation. Tucked down into the shadiest and steepest zones right below our oak and brush thickets are scattered large groves of dark-green conifers, not what you'd expect above Palm Springs. Then great vertical bands of brush and grasses, dotted with yucca and rocky outcroppings, alternate with oak forests choked with decades of dead brush and branches. All these types of vegetation cling precariously to cliffs or steep mountain walls, especially the conifer forests that sweep all the way up one ravine to touch the ridge for a distance of a quarter mile or so.

The fire still lies like a sleeping monster. Its wind-fed appetite may be temporarily satiated, but its potential to awaken for another snorting, rampaging run the rest of the way up the mountain seems only a matter of "when," not "if."

From one particularly good viewpoint I can see literally miles of denuded mountainside down across the curving range ahead of us. It appears that some of the hottest spots still burning in the temporary lull lie directly beneath the fir forests that we're cutting our way toward now. But there's so much work to do cutting roots and chopping out the cork-like matting of duff that I can only glance occasionally toward the smoke.

The winds die to almost nothing. Everyone's already feeling the strain of hours-long exertion in the direct sun, and even though our hard hats offer slight protection from its blistering rays, they also hold in our body heat along with our long-sleeved fireshirts and the daypacks we wear on our backs.

Water is priceless in this heat on a hot-line shift, and each of us tries to drink sparingly. What we carry is all we may see for a long time, but cracked dry lips and raspy throats make it nearly impossible to just sip from our now lukewarm canteens.

Suddenly, one of the other crew's sawyers cuts his way into an unseen hive of meatbees, or yellow jackets. Brave and fearless firefighters flee, scrambling wildly in all directions as curses and yells identify the unlucky victims too close to escape the bees' onslaught. Less than a minute after the first sting, almost everyone's regained his composure and "Kid," one of our youngest sawyers, disdainfully walks right back past the bees' nest to resume cutting.

A little more reluctantly, the rest of our saw teams gingerly work their way around the hive, widening the line a bit downslope to compensate for the uncut fuel right next to the bees. The rest of us, fortunate to be far behind the saws during the attack, are surprised to see one of the other crew's squad bosses running up through our midst carrying the first aid kit.

Denny moves up to find out the problem and soon walks back to announce that one of the other crew's sawyers who got stung is allergic to bees. Even as he tells us, we can hear a message over the radio on his hip calling for a helicopter. Obviously, someone on his crew thinks this is critical.

Ten minutes later, our line-building squad approaches a patch of large oaks rising above the surrounding brush that borders the wide canopy cut. Beneath the tree's welcome shade sits the victim, hard hat off and a wet bandana draped across his forehead.

His face is quite puffy and he's obviously having difficulty catching his breath. Despite the seriousness of his situation (waiting here with medical attention at least thirty minutes away), two of his crewmates are teasing him mercilessly. "Some people will do anything to get out of work" seems to be the commonest joke, but I notice that his squad boss is surreptitiously glancing at his watch and timing his breath rate.

By the time the copter approaches, the saw teams have widened a small clearing on the ridgeline into an acceptable helispot. We watch grimly as the sting victim shakily waves a "thumbs up" gesture to his buddies as he clambers up into the ship. Will he make it to the hospital in time?

The ship's soon just a distant speck and all of us lean a little quieter into our work. Here's a fire that's more than 10,000 acres in size already, but a tiny bee might be what takes a life.

Late afternoon sunshine beats unforgivingly on our bent-over bodies, as hour after hour we slash our way up to the rocky knob. Scraping and cutting becomes automatic, almost like breathing, and I work in a state of steady monotony. By early evening, the saws are close to tying in, and looking up, I can see the Big Guy and two of the other "supes" sitting with the other overhead bosses in a cluster on the highest rock, watching our approach. They

Dry Falls

laugh a lot, a soundless laugh unheard above the whine of cutting chainsaws, but I can see them joking and nodding approvingly.

Back aching, arms wooden, and spirit numb, I finally swing my last cuts and tie into the rocky bluff. Following Larry and Dan, I join the sawyers who lie in rag-doll positions scattered here and there on the bare patches of soft dirt amidst the countless big rocks. We all rest in near silence, except for Kid and his puller, who argue relentlessly for fifteen minutes over some meaningless aspect of their day's work.

Paul finally comes down from the overhead rock to tell Kid to shut up, followed by the Big Guy, who rousts us up to follow him downslope through the brush. We slip down a freshly cut narrow path below the long rock bluff to suddenly emerge on a wide, well-maintained trail 200 feet or so off the ridgetop. Leaving Paul to bring us along after a short food break, the Big Guy moves off ahead along the trail, disappearing almost immediately into the fir forest we've been cutting toward for so long. Above us, the forest bumps up against the rock outcropping, but I remember seeing this same section of conifers reaching all the way to the ridgetop up ahead of us on the other side of the knob.

Paul checks us out for gas and oil as well as the condition of our saws and tools. We listen to his latest description of the fire's size at 16,000 acres. Apparently while the fire lay dormant here most of the morning with the winds pushing downslope, it still burned strongly to the east far below us, as well as rushing upslope in the few big runs it made up ahead of us. At last report, there are more than thirty handcrews encircling the fire, with almost the entire ten miles of the most dangerous upper edge of fireline being built by Hotshot crews. Obviously, we'll be here awhile.

We gulp down squished white-bread and ham sandwiches, apples, candy bars, and cans of juices, the standard fare in every bag lunch we ever get on the fireline. Then, still a little hungry and even more tired, we quickly sharpen our tools and saws in the last bit of light before dark. Then we put our headlamps on our hard hats and regroup.

Paul lines us out slowly, knowing how tired we are. Immediately as we enter the forest, we begin to see our first close signs of fire for the day—burning logs, smouldering branches, and forest litter. We round two bends just in time to watch a big run of fire rush up a very narrow ravine, then spread out above the trail ahead of us to disappear into the upper forest in a rush of burning foliage and crackling noise.

One of the sawyers jokes, "I'll bet that made the Big Guy run," but not many of us laugh. A minute later, two more slopes ignite, lit by upslope breezes pushing fire up out of the smouldering brushfields far below us.

Paul radios the Big Guy, then moves us all forward again into the still-hot forest. A big burning limb clunks, crashes, and finally disintegrates as it lands right above Paul and a sawyer. Everyone tightens up, forcing alertness in this eerie march through the firelit night. Most of the fire's runs have been erratic, leaving whole sections of trees untouched and others totally involved. We soon reach an especially steep hillside, so steep that the bank above the trail rises nearly vertical far above our heads. Here, and for a long ways ahead along the trail, the slope beneath us is a cliff, where a misstep could send us sliding to our death.

Right at one of the vertical gullies, where the trail crosses a steep chute, a great burning oak tree has just fallen over right above the narrow crumbling path, radiating such tremendous heat that despite shovelful after shovelful of dirt thrown on it, there's no way we can cross. Paul finally waves us back to wait a few minutes for it to cool down. Exhausted, we slump against the hillside or sit on the trail's edge where it skirts the cliff, our feet dangling in space far above the rocky slopes below us.

This is one of those unimaginable times in firefighting, one which is hard to explain to loved ones at home later in the security and bright lights of a warm house. The night air is mostly still now, moving just slightly upslope as the fire's heat and smoke rise. For as far as we can see to the north of us in the darkness, pinpoints of light mark thousands and thousands of burning stumps, logs, and bushes. Even in desolate areas burned more than a day before, we can still look across the dark mountainsides and see myriad specks of flickering light, slightly different in color and appearance than the many blinking stars visible through the drifting smoke higher up.

Far, far down and out at the edge of our mountains, we can see the colorful lights of Palm Springs, floating in the darkness with all their neon and neat rows of streetlights and shopping centers. Then close at hand, thirty or forty big conifers burn high up along their trunks. Most of them are live trees, dying slowly, but a few giant dead snags burn fiercely. These snags are the real show, glowing and sputtering like brilliant fireworks displays right around and below us. Occasionally, a huge chunk of some burning tree will suddenly fall, crashing down and down out of sight into the forest below us.

Now that we've stopped, our dripping, sweaty clothes start to chill us, especially as the cool night breeze increases. Most of us eat the last of whatever munchies we've saved until now, but the meager carbohydrates and even sweatshirts can't warm us; our fireclothes are just too wet and ice-cold.

For almost an hour we wait in the silent darkness, waiting for the burning oak to cool down. Larry jerks out of a drowsy half-sleep next to me when a big rock loosened by burning up above us suddenly bounces down out of the forest. It's not even close to us, but other firefighters farther over on the trail scatter until the boulder crashes on past to bound on down the mountain into the darkness. Soon after, the radios crackle as the Big Guy wants to know why we haven't caught up yet. Paul explains about the tree and the airways return to silence.

Ten minutes later, however, we hear the Big Guy yelling across the gully, and half asleep, we all line out and file toward the still-burning log. Under the Big Guy's direction, six men take turns throwing dirt at the incredible bed of red-hot coals that the big log is slowly changing into, but the radiant heat is still too hot to even run past safely.

More dirt, then more, and then the word is passed: "Line out and cross!" Larry and Ann are the first to complain, pointing out to Denny quite rightly that a slip in the middle part of the fifty-foot crossing will send us sliding straight off the cliff. I notice also that all the dirt throwing has knocked lots of hot embers all over the narrow path. To cross through such intense heat on such shaky footing seems risky, to say the least. I can even hear Paul argue once, then give in to the Big Guy.

Reluctantly, we all move forward, one at a time, each one ducking his head, shielding his face with his arm, and dashing with his eyes closed past the blazing oven of a log. All goes well until Dan's turn. He darts forward along the trail but stumbles in the loose, burning coals right below the log. For a brief moment we all fear he's going over the edge, but he stumbles forward and lunges past the greatest heat to be pulled forward by Kid to safety.

Unnerved, Ann and Larry hang back, so I move forward, plan my route carefully, and rush across. The heat's too much to take, but only for a few seconds, then I'm through into the cool darkness on the other side.

The last few firefighters hurry across as we regroup and follow the Big Guy along the trail, around the mountain, then up and up to the chill wind of a barren boulder-strewn ridgeline. We huddle around tiny warming fires, waiting in shaking, teeth-chattering clusters, knowing that soon the other crew will join us to begin our next section of fireline. Long before any warmth has begun to dry our freezing clothes we watch the other crew drift ghostily out of the darkness. It's work time again.

Just as before, we line out to cut fireline, but this time our two crews take different sections. Together we must follow and build line along the burning or dead fire edge of the long run that roared up to this ridgetop during the heat of the afternoon. We work mechanically, half dazed as midnight approaches and we're still slamming our tools into the smouldering brush, chopping at deep duff and leaf-covered rock that sends sparks flying off into the darkness.

At two o'clock the dull monotony of bent-over strenuous labor in the vast empty-feeling mountain darkness is broken. Someone suddenly radios for help and our squad rushes uphill to help battle an unexpected hot run through a steep hillside of unburned down logs and oily ground covers. Instead of feeling cold, we've switched back to sweating as we fight the spreading fire across the mountainside for two hours. We even lose it once, watching in horror as it escapes our efforts and runs right over the top of the ridge to slow just enough on the other side for us to quickly hook it and bring our line around it and back to this slope again.

By dawn we've got a good handle on it, and as the first faint red glow blushes on the eastern horizon, we stagger up to our goal, another rocky outcropping. Like zombies, everyone slumps

or sprawls wherever the rock is moderately bare. No one talks.
Thank God the night is over.

Helicopters are the only sources of supplies on many Hotshot assignments. *Author's photo*

four

A Wilderness Fireline

I roll over slowly to catch the first warming rays of the rising, blood-red sun. Bodies lie motionless all across the clearing, some twisted in such grotesque positions that it seems impossible for anyone to rest that way comfortably. We've been awake more than twenty-five hours, most of that spent working harder than most people can imagine. From where we lie we also have a view that few could ever dream of, and even fewer have ever seen from this particular rocky knob.

Ann hoarsely whispers, "Wow!" and her awe is shared by most of us. A colorful spectacle of steadily changing reds and oranges covers the sky from north to south all the way along the eastern horizon. The drifting smoke from our mammoth wildfire has blown eastward, enhancing and reflecting the rising sun's light into a bright display of golden light rays shining up through dark, colorful bands of backlit smoke.

From our mountaintop perch we can see far out into the desert, stark and surreal in the indistinct blending of this sunrise lighting. Looking down at Palm Springs, darker than the surrounding desert, we can still see its sprawling subdivisions and golf courses separated by rows of orange and yellow streetlights that string out like glowing beads into rigid trails of lights. All the mountain slopes in between lie charred and devoured, littered by the blackened twisting silhouettes of thousands and thousands of stumps and partially consumed brush trunks.

To the north along our mountain, orange and brown patches of scorched but unburned conifers and oaks stand out distinctly from the green or blackened forest on either side. And here, high up on the top of this rocky ridge, stand scattered healthy sentinels, windswept, bent-over pines and live oaks somehow thriving with roots gripping deep into the solid rock of the mountain itself.

Toward the southeast we can see over our long scar of a fireline into a lush, high-elevation valley—the one where we left the school buses to begin the long ascent up the back side of this range of hills. And far, far toward the west, I can barely make out a thin band of dark clouds that may be coastal fog pushing inland over millions of people just waking up to begin their normal day.

"Let's go! Line it out!" shout Paul and the Big Guy, walking amongst the crew with firm nudges and shakings of shoulders. Bitching and moaning, everyone lines out . . . a sorry-looking bunch with shirttails hanging out, sleeves torn, and faces grimy and puffy. We march slowly down off the first peak of the knob, through a maze of boulders, and then up to the other rock pile of a summit. At the edge of a great dropoff, so steep that boulders right beneath our summit block us from getting any view of the ridge far below us, we meet up with the Plumas Hotshot crew.

The Big Guy holds us up for a few minutes, long enough to talk to the Plumas superintendent and get the details of how the fireline has progressed from this point back to the north. Apparently Plumas bumped around us during the last of our night shift to pick up line construction here, but dead tired, waited for morning sunlight.

The Plumas boss is new on their crew and from the sullen looks on a few faces, he's obviously dealing with discontent from his tired and unenthusiastic crew. The Big Guy walks our crew on past the Plumas Hotshots, off the west side of the rocky outcropping, and over to the very brink of the dropoff. Then he gestures, and we all close in to hear, but moving with care. One wrong step and we'll fall out of sight.

"Hey, those guys are arguing it's too dangerous to go down here and their new superintendent's trying to convince them to go. Can you believe that? Anybody ever questions me and they'll be looking for a new job! Any questions? Okay, let's go!"

There are a lot of questions, of course, but none of them spoken, just winks and exchanged looks and lots of muttered slurs about someone's lack of mental capabilities. Yet we each know that no one—no one—can actually make us go where any of us doesn't want to go. But one by one, our crew steps out into space and drops down a few feet to land on the next boulder, then the next, swinging out of sight around the rock bluff.

The descent is nowhere as difficult as anticipated, even for the sawyers who struggle to balance heavy saws while carefully placing their Vibram-soled boots on the narrow, precarious footing down through the steep boulder field. Even before I reach the bottom, our saws are cutting again, and as casually as if we'd never stopped, we all move out in fire tool order, cutting line from the bluff along the next ridgeline. To say hunger gnaws at my stomach isn't quite true; I'm mostly past the physical discomfort of hunger — that was last night. Now I'm just feeling groggy and light-headed from our long shift.

The sun feels good, though, and we all seem to have gotten a second wind, for everyone except for a couple of our laziest workers still manages to keep up a fairly good pace as we cut line through an open pine forest. The Plumas crew, having overcome their reluctance to drop off the steep slope, appear from behind us and leapfrogging ahead, really help us pick up the pace. Lassen and another Hotshot crew pass downslope from us on the newly visible trail (which someone explains is part of the Pacific Crest trail system that stretches from Mexico to Canada), waving and shouting as they head off toward a heli-spot somewhere ahead for breakfast.

The idea of a breakfast perks everyone up, and soon even Plumas has disappeared toward the well-earned promise of food and rest. We work on along for another half hour until direct contact with a sector boss convinces the Big Guy to also head us toward the heli-spot.

Heli-spots, however, on this fire are few and far between, so even after we clamber down the mountain to pick up the hiking trail, we still march for over an hour before we finally, at long-aching-legs-last, plod exhausted into a nice pine forest clearing below the heli-spot. Dreams are made to be broken, or something like that, for all the food is gone except for a few cans of fruit juice, a box of yesterday's squished white-bread-and-bologna sandwiches, and a box of apples. To say everyone's pissed off is an understatement. We're furious, but the spike camp manager shrugs and starts listing all the problems of the helicopters and supply efforts.

Famished, we eat the meager leftovers, walk up to our

designated sleeping area under some large pines, and spread our gear. Within minutes, despite helicopters flying in crews to the saddle in the ridge just upslope, I'm gone, totally oblivious to sun, flies, mosquitos, or even Denny's snoring. For the moment at least, this Hotshot parade has halted in the shade, and once again, sleep is a wonderful thing.

I wake up dripping with sweat. The sun has shifted during the five hours I slept, and now I'm lying, drenched in sweat, directly in the full summer sunshine of Southern California. Still trying to keep half asleep, I push myself up, grab my paper sleeping bag, and flop down on it again deep in the shade at the base of two big pines. Just as I snuggle down into a comfortable position and start to shut my eyes again, I catch a glimpse of the Big Guy cruising upslope toward our slumbering crew.

Reaching Paul, he wakes him, gives him our schedule, and heads back down to the piles of orange canteens, garbage, and cardboard boxes that mark the center of spike camp. Paul nudges Frank, the squad boss for the saw crews, who yawns and grudgingly begins to wake up all those close to him. Like a ripple spreading, heads pop up in slow confusion, staring sleepy-eyed and bewildered as tired brains try to perceive just where we are and what's up.

"The Big Guy wants us first in line to eat so we can get out of camp before some overhead boss holds us here all night," explains Paul. "After that first long shift, we may as well get all the overtime we can."

Everyone grumbles and yet agrees at the same time. We'd love to sleep for ten hours, but we'll make much more money today if we hurry back out to the fireline. Ten minutes later we're all lined out, waiting for the hot food that's been promised on the next incoming helicopter. Sure enough, right on schedule, an S-212 beats its way up the mountain from the valley far below us, trailing a netted sling-load of water and containers of hot food.

We all kiddingly jostle for position as the spike manager and his assistants stack up and open the food containers. It's been a day and a half since we've eaten a real meal and we're ready. But bad luck strikes again. Somehow there's been a mix-up at base camp, and all that's been flown in are containers of lukewarm cooked mixed vegetables and one big container of mashed potatoes. There's no meat, no milk, no bread, and no dessert.

To say that the Big Guy is livid would be an understatement. He's wagging his finger right in the food manager's face, telling him exactly what message to pass on to those S.O.B.'s back at base camp. Then, to cap off everything, we hear base camp apologize over the radio but explain that there may be a two- or three-hour delay before the rest of the food gets here, if it even does before dusk ends all flights.

Before we really know what's happening we're all struggling into our web gear and daypacks, hoisting our tools and heading off behind an irate Big Guy, who's in no mood to hear any suggestions that we at least eat what's here. Instead, we march back along the old trail, reflecting on the circumstances that turned the promise of a delicious hot breakfast and a hot afternoon meal into one stale sandwich and an apple. At least we've gotten fresh water.

Retracing our earlier route, we return to the uncompleted section of fireline upslope from the old trail. Lacking enthusiasm for anything except complaining, we nevertheless line out and begin cutting the patches of brush and ground covers along the narrow ridge. The Big Guy and Paul tie in with a sector boss who's full of nothing but praise for our performance the night before and sympathizes with our hassles at spike camp. He readily agrees to the Big Guy's suggestion that we hold our previous night's line after tying in the piece we're working on now, and the good news quickly passes down our line.

Just before sunset we reach an already completed section of ridgetop fireline, tying in the last unlined section for miles. We break for ten minutes, then drop downslope, pick up the old trail, and follow it south back to our first day's arduous efforts. Joined by the well-rested Lassen crew, just returning from the spike camp, we spread the two crews out for a half mile along the ridge.

Almost forgotten since our late-night battle the night before, the fire has mostly died out far downslope all along this long stretch of mountain, leaving only a few smouldering hot spots to threaten us with slow-rising plumes of drifting smoke. But despite the quiet, we can't count on the thick, oak-studded brushfield below us staying quiet, since countless small fires still burn slowly in its midst. Orders come down again. "Widen the fireline by another ten feet!" Ah, there's no rest yet.

Saws whine to life, everyone falls into place, but there's no adrenaline to spur our efforts as we slash our way slowly along the ridge. Yet despite hunger, exhaustion, and frustration, most of the crew keeps at it. Admittedly the whining from various crew members definitely rivals the whine of our chainsaws blaring up ahead, but even while they bitch, the crew works.

Darkness finally forces us to halt long enough to put on headlamps, but not until almost midnight do we finally reach the knob with a widened fireline. As the overhead drift up to their lookout perch on the top of the knob, the rest of us spread back down the line, hopeful that at last we'll actually get to "hold" line rather than construct it. The Lassen crew disappears over the knob, their headlamps vanishing one by one as they move out of sight over to the next ridgetop.

Paul drifts down our line, talking a few moments to each pair of partners (we break up into pairs for safety as well as companionship). He flicks his headlamp off as he squats down beside Larry and me, blending back into the dark cover of the night.

"The sector boss has headed in, the Big Guy's shooting the bull with Kid, and it looks like we're just going to hold this line the rest of the night. The Big Guy says if all's still quiet at two o'clock, we'll build a couple of warming fires and cluster till dawn. So stay alert, okay?" Then Paul laughs and Larry and I join in. We all know just how alert we feel. Even the freezing cold breeze blowing in across the ridge won't be able to keep us awake too long.

Paul slips out of his daypack and digs around inside. "You guys hungry?" he asks, offering us what looks in the dim starlight to be a bag of roasted nuts. His question needs no reply. Larry springs to life instantly and gratefully takes a handful, repeating his thanks over and over. "Wow! This is great! I'm starved!" I'm glad too and slowly savor each bite.

Once Paul's gone we seem all alone on this big mountain. No headlamps light the wide fireline and only the sound of the dry whistle of the wind makes any noise. Two o'clock finally arrives and all of us head up to huddle around two small campfires built out in the middle of our highway-wide fireline.

"Stay awake!" admonishes the Big Guy, who immediately takes over the warmest, smoke-free half of the largest fire and begins to joke with a few sawyers. Six others crowd in as close as they can get to the other fire, leaving ten of us standing cold and disgusted in the windy darkness.

Denny suddenly taps me on the shoulder and points down the ridge to the shelter of a jumbled mass of sharp rocks sticking up from one side of the denuded fireline. "You guys go ahead and sleep down there," he whispers. "I'll yell if there's any problems."

I can't honestly say that a bed of loose rocks on a windy ridge is wonderful, but with a little effort we all manage to find a suitable spot. Each of us carries a space blanket, a light aluminum foil-backed sheet that blocks the wind and protects us from the moist late-night dew. Whoever designed the space blanket, however, must have carefully calculated to make sure that each space blanket is just a little too small to actually wrap fully around an adult.

Every time we move, the stiff space blankets rustle loudly, and for at least a half hour the steady rustling of our aluminum blankets dominates our little sanctuary from the wind. Finally, even though we still haven't managed to protect both our legs and our heads, we've made personal decisions on which one to sacrifice and have settled into some sort of final position.

Denny's snoring leads me to believe he's not going to make much of a look-out for the rest of us, but the only action on the ridge is the loud voice of the Big Guy retelling fire stories up at his fire. I wake up once, sometime before dawn, with brilliant sparkling stars absolutely filling the sky from horizon to horizon. Numbly, I wiggle one hip a little to relieve the prodding discomfort of an especially sharp rock and notice that all's quiet, even up at the Big Guy's fire.

All's quiet, that is, except for Denny's loud snoring, and I fall back asleep to his rhythmical snorting and breathing, delivered melodiously on the ice-cold pre-dawn breeze.

I don't recommend waking up face down in the dirt between rocks; at least I suggest working up to the experience gradually. But despite swollen eyes and a filthy face, I must admit it's another beautiful morning.

"Come on! Get up!" bellows Kid, doing his best to be obnoxious at this early hour of the morning. Even a little sleep, short as it was, was great. Fireline safety frowns on ever sleeping on a fireline. Reality, of course, doesn't always match safety plans. Hotshots must be in top physical shape, able to endure beyond normal limits, but even Hotshots get exhausted.

"Pancakes, bacon, and eggs!" yells Kid, teasing some of our leading complainers who are already moaning for food. "Sweet rolls, hash browns, and orange juice!" Kid goes on, only to be cut short by a barrage of rocks whistling over his head.

We spread back along the ridge, checking intently for any signs of flare-ups or spot fires during the night. Nothing is visible anywhere near our fireline and even the drifting smoke from the main fire far down the ridge has almost completely faded.

The early-morning hours pass quietly as we bask in the welcome warmth of the sun. By midmorning a worried division boss hurries up the fireline, trying to find out how we're surviving after two days and two nights. I imagine we look pretty hardcore, but just the couple hours of shut-eye we managed before dawn seem to have invigorated all of us. Besides, if there's one thing a Hotshot crew can really do well, it's hold line. Only hunger keeps this easy shift from being quite pleasant.

After a short discussion with the Big Guy, the division boss starts talking on his radio. Within five minutes, we can see the Lassen Hotshots filing one by one around the knob toward us. I can tell by just looking at them that they caught a little shut-eye themselves.

Paul and the Big Guy collect us two by two as they come down the ridge until we're all lined out behind Lassen and moving down to the old trail, headed south. We follow the trail for a mile or so on the east side of the ridge before swinging down through a low saddle where the trail drops steeply to the west. Far below we can make out a ranch with Hotshot crew buses parked in rows. Almost before we know it we're down to the main buildings, joining another five or six Hotshot crews who've also been brought down off the fireline.

A Wilderness Fireline

It's a Hotshot reunion with back-slapping, wild bursts of laughter, and lots of grunted nods of acknowledgment to the many firefighters we know only by their familiar faces. A Mendocino crewman comes over to tell Denny that this ranch supposedly used to belong to Timothy Leary, the LSD guru. The key ingredient in his tale is the whispered assurance that the large ranch pond still contains the undiscovered remains of victims of drug overdoses.

We all drink in every word without scoffing; he makes the tale sound believable and it adds interest to our wait for a ride to the nearby firecamp, a smaller version of the main base camp we originally stayed at in Palm Springs. Soon bright yellow school buses bounce stiffly and loudly over the bad dirt road leading into the ranch and one after another they fill and go back again, loaded to the brim with men and equipment.

Most of the bus drivers are men, but we have a spry older lady who handles the shifting and big steering wheel as if she's been driving for twenty years. It must be quite a contrast to her usual busload of elementary students—twenty stinking, filthy, loud-talking men all wired up at the thought of good food, cots, and sleeping bags, maybe even a beer or two if we get some free time near a town. Some of the ruder guys, totally lacking in the least respect for either the bus driver or good manners, spit their spent wads of chewing tobacco right on the floor of the bus, while a couple of the others talk loud enough that every swear word that's basic to their limited language floats up to the front of the bus and the burning ears of the lady driver.

She looks like she's really steaming by the time we finally reach the new firecamp and simply clenches her teeth in disgust as our crew parades off the bus without even a polite "Thanks." But all eyes are focused on one thing—and yes, that really is a food line already forming in front of the kitchen area.

We eat a full hearty breakfast, even though it's nearer noon, then set up our own little camp unit complete with cots, paper sleeping bags, and a couple of long, bright-yellow tarps that we string up high enough to make a large shaded sleeping area. Despite a crush of bodies, we all wait our turn to crowd into the portable showers.

If you can remember the indignity of your high school locker room showers, you'll get a hint of the mixed agony and joy we feel heading into these portable showers. We all need them; even the most animalistic of us acknowledges the urgent need for bathing after days of sweat-drenched labor. But few gambles are as risky in life as betting on the appropriate warmth of a portable outdoor shower unit. Ready for the worst, we trudge toward the waiting line that forms outside the canvas-walled shower area, most normal conversation overwhelmed by the loud generator that's powering the pressure system, heating system, or lighting if it's at night.

When it's finally our turn, we walk in on narrow wooden slats laid out to keep you from having to stand directly in the mud, reach up and switch on the little nozzle that squirts a small stream of water down in its own individual pattern. Some showerheads blast with one tiny straight stream of high pressure that feels like it's going to bore right through your skin. Then a few showerheads over, there'll be one that shoots out such a fine mist that even standing right beneath it you never really seem to get wet.

The same variety applies to the temperature. Long forgotten is the luxury of actually being able to adjust the heat. Rather, to have any heat at all after 200 firefighters have swarmed the shower unit within a half-hour period is a stroke of good fortune not usually expected. Even on a hot day, ice-cold water may take some especially strong willpower to withstand for more than a few minutes. And without rough scrubpads, it takes time to even start to remove our individual portions of mountainside and ashes we've unintentionally brought along. Occasionally, the tanks of shower water get heated so hot that the showers turn into a fiery ordeal of endurance. But hot or cold, the crowded showers resemble a war zone like $M*A*S*H$, with army-green canvas and army surplus water tanks intermingled with naked bodies and piles of paper-cloth towels.

Eventually our crew filters back to our sleeping area. We're all really tired, but it's hard to fall asleep with temperatures over 100 degrees and the bright sunshine glaring dully through the tarps. We lie on our cots, wearing only shorts or cut-offs, swatting continually at the flies. Nevertheless, half the crew's asleep and the rest of us are close when Paul approaches and loudly announces, "Get dressed, everyone! We're heading to Palm Springs! They're going to feed and entertain us!"

No matter where they eat, or what's served, Hotshot crews eat like there's no tomorrow, for, in terms of meals, often there isn't.
Courtesy U.S. Forest Service

five

A Palm Springs Bash

I'm sure we look quite impressive, decked out in our blue Hotshot t-shirts and red ballcaps as we wait to board one of the remaining school buses for the ride down to Palm Springs. Already, most of the other crews have taken off, but Lassen, Plumas, and a couple of others also took a few minutes to get it together.

Our bus driver this time is a middle-aged black man with sweat dripping down from his frizzy Afro haircut onto his ebony forehead. He eyes us suspiciously as we push up into the bus, laughing as we squeeze long legs into narrow spaces designed for school kids. He pulls out onto the narrow-laned highway with a lurch, either ignoring or not hearing a casual question about how long it will take to get there.

We settle back comfortably into our seats, yawning and hoping to at least catch a few winks on the long ride to town. Sitting just across and back from the driver, I gradually notice how nervous he's acting. His knuckles are clenched super-tight on the wheel and his shifting is rough, to say the least.

No one else seems to notice until we're about five miles down the road. He downshifts for a long, slow curve and grinds the gears badly, swerving as his concentration slips from the road ahead to the gearshift. A couple of our perpetual comedians laugh and make some teasing comments, but instead of laughing back he tightens up still more and continues weaving down the road.

First one, then two more school buses pass us on the long straightaways, roaring off toward Palm Springs while we grind along at thirty or thirty-five miles per hour. He drifts out of our lane again as we enter the first major curves of the long series of switchbacks through the deep, narrow canyon section that stretches more than ten miles down toward the flatter desert and Palm Springs.

We sway first one way, then to the other, and now I'm sure

that we've got something to worry about. We slip past a couple of trucks heading uphill the other way with no problem, then suddenly this guy's got the bus in the middle of both lanes as we enter a blind curve, and now it's my turn to clench knuckles and grip the seat. By now most of the crew have taken more than a passing interest in our driver's driving; even the Big Guy yells out something like, "Hey! Let's keep in our own lane!" But the driver only starts muttering out loud, sweat running down his face now worse than ever.

He slows way down for the next curves, but soon a big truck behind us moves right up on our tail, threatening us with his closeness. Our driver more or less follows our lane of the road, but with no shoulder through most of the steepest sections, he still takes his half out of the middle, narrowly missing every other car or truck climbing the mountain.

I can't hold back anymore and ask, "Is there something the matter with this bus?"

He turns to look at me straight in the eye for one long moment, then jerks his head back to worry out a route along the road ahead. "Damn buses," he finally spits out. "They've got me up here on these crazy roads and I've only been driving a week. Never been out of Palm Springs before. Damn these buses!"

Only those of us close to the front hear his exasperated confession, but we look at each other with rolling eyes and shaking heads. "Great!" whispers Nell from behind me. "New headline! Hotshot crew dies in fiery school bus crash! We'll make the news yet!"

"Shut up!" I mock-scold back in a whisper, but I couldn't agree more. This guy is an accident in the making.

Half-in, half-out of our lane, we weave drunkenly down the twisting road as the night sky darkens. Just when it appears our driver may actually deliver us in one piece to our destination, he overcompensates entering a long curving bend and the bus swings all the way across the road, barely gripping the far edge of the opposite lane before lurching back across to safety. Guys start yelling and cursing from the back; others who've somehow slept the whole trip wake up confused.

Those of us at the front just take deep breaths of relief and welcome the sight of the lights of Palm Springs just ahead of us.

So much for the entertainment part of our promised evening!

Past the fancy restaurants, back past Bob Hope's supposed home on the hill, past the security guards and toll gates leading to the homes of the rich or famous, our school bus finally grinds to a halt outside the grounds of Palm Springs's extravagant high school and community sports stadium.

Boy, are we glad to get off that bus! Grateful just to have survived the ride, we line out behind Paul and march between school buildings across to the brightly lit football field. Loudspeakers blare out someone's enthusiastic solo on guitar as a California Division of Forestry captain passes us and calls out, "Hey, you guys are late! Better hurry before all the food's gone!" He winks at our grins and points toward the gate at the end of the field.

Kid shouts out loud enough for everyone to hear, "Bring on that food! I can eat a whole turkey myself!" Larry and Dan argue between themselves as to whether a buffet includes dessert or not, while everyone's either guessing or hoping that some of the celebrities turn out to be their personal favorites.

Great banners stretch across the far side of the field declaring "THANKS FIREFIGHTERS!" in big letters. Huge lights flood the field in noontime brightness and even as we file through the gate we can see that the field is jammed with hundreds and hundreds of people. Centered on the far side of the track, there's a large stage where musicians are scrambling about, setting up equipment and tuning up for the next songs. Everyone's standing or sitting in a big cluster that curves in a sort of arc facing the stage, and back behind the crowd runs the longest string of tables I've ever seen. Each table butts end to end against the next, and each is covered with a bright tablecloth featuring a huge centerpiece of flowers. Pretty fancy, to say the least.

We seem to be the last crew here, for paper plates and styrofoam cups stuff garbage bags to the limit all amidst the crowd. Puzzled, we file along the twenty- or thirty-long string of tables, noticing that none of them look like they've had any food on them.

Near midfield we run into the back of the food line and surprise, surprise, there isn't any. "Looks like there's more of you boys than we planned on," consoles an elegantly dressed matron, "But don't worry, we've got more chili and salad coming any minute now."

We look up the tables and sure enough, there's not much on this great string of tables except some kettles of punch and lots of empty boxes of catered food, obviously donated by local restaurants. We're not alone in our wait, for Mendocino and another Hotshot crew stand ahead of us in line, eyes glazed in the same sort of shell-shock that we feel.

One of the Lassen Hotshots waves to Nell from over in the crowd and gets up to come over. "Don't feel bad," he says. "We got here almost an hour ago and all they had then was chili, bread, and salad. We heard that all the good food went to the camp slugs." (Firefighters' derogatory term for those necessary, but less physically active workers around firecamps.)

We wait and wait, stomachs growling and hopes fading. A sleek Mercedes finally purrs down the sidelines and another well-dressed lady emerges, this time followed by friends carrying tubs of shredded lettuce and trays of warm French bread. "Help yourself, boys, don't be shy," encourages the first lady, and slowly our crews follow the line up to the lettuce and bread.

"Here, you must be hungry!" waves another local entrepreneur, and before we know it he's stacking two ice cream bars on everyone's plate. Plain lettuce, French bread, and ice cream bars . . .? Larry looks back at me with one of those "Tell me this isn't happening" looks as we grab a cup of punch and search for a seat on the grassy field.

"Ay-yay-yay-yay!" whoops a Mexican-music trio on stage, trilling and dancing at a fast pace through a couple of totally unheard-of melodies. Denny sits down beside us and comments, "I'm sure their intentions are good, but those guys are lousy." Perhaps "lousy" is a little too critical. It might be closer to the truth to say that their talents have not quite reached perfection yet.

One of Denny's friends from the Ojai Hotshots joins us, wincing comically as one of the singers stretches for a high note. Denny quietly asks him, "Did we miss Bob Hope and the starlets?" His friend shakes his head. "The same three groups have been taking turns—our Mexican wonders, an old guy who tells jokes, and three elderly matrons who play violin, piano, and cello. Believe it or not, they're actually pretty good. But nope, the only stars here tonight are those overhead."

We all laugh as another illusion pops abruptly. There's something about eating shredded lettuce, French bread, and rapidly melting ice cream that seems illogical, but funny, and we might as well make the best of our downhill-sliding night.

For two hours we mingle on the field, stretched out most of the time on the grass, listening to the good-hearted performances. A pretty newscaster lady, accompanied by a dapper assistant and two cameramen, circulates through the crowd, stopping here and there to film a few seconds of someone eating an ice cream bar or a group clapping enthusiastically when the Mexican-music singers declare this is their last song of the set.

The news team brings a tall, good-looking CDF officer in starched shirt, shiny shoes, and pressed pants into the midst of the crowd to film an interview with the stage behind him in the background. We can hear him assure the lady that "We've got a real handle on the fire now" and "It took a great effort on the part of everyone, but the worst is over." She seems to eat up every word, but all we can do is laugh. It's unlikely that this guy's been up on the mountain where most of the action's taken place since the fire burned mostly uphill away from the town. We also know that there are still miles of uncontained fire perimeter on the southwest section of the fire, so maybe the worst isn't over. But it all makes for good news coverage.

At last midnight arrives and the "master-of-ceremonies" thanks us all again for "saving" Palm Springs. We all trickle back through the maze of school buildings to grab paper sleeping bags and cluster as a crew out in the middle of the large grass practice field where we slept the night we first flew in here. By twelve-thirty everyone's settling down, our crew as well as the ten to fifteen other crews spread around the huge field. I'm almost asleep when Paul comes over and whispers, "Look down there at the end of the field! Is that for real or are my eyes playing tricks?"

I look and maybe we're both crazy, for there, munching away completely carefree is a giant bighorn sheep with great curving horns and muscles that stand out even in the dim light from a hundred yards away. Any other time and I'd be up and investigating, but other than to lie there watching, I'm just too burnt out. Besides, we've already heard that they're waking us up at four-thirty and it's almost one A.M. now.

I'm just about ready to quit watching the ram when a bunch of guys from another crew decide to get close to the bighorn and catch it. To say the idea is stupid is a great understatement, for immediately the ram bluff-charges toward them, scattering "macho" men like mice. Then a pickup truck drives onto the field with some camp personnel who vainly try to herd the sheep out through one of the three small gates off the field.

Suddenly the ram charges for real and crashes into the door of the pickup with a loud bang that reverberates over the field and wakes anyone who wasn't already awake. Not content with a badly dented door, the camp guys clap their hands, wave towels, and finally drive the ram past us into a dead-end alley between two classrooms. Four guys quickly block off the entrance with a bunch of garbage cans and cheer, "We've got him!"

Approximately ten seconds later the bighorn hits the stacked garbage cans full force, battering them apart with such velocity that lids and cans seem to literally explode out from the alley. The would-be pursuers dash frantically in all directions and only a heads-down blind charge on the part of the wildly running ram keeps him from easily obliterating one or two of the slowest ones.

Within minutes a big camp boss is standing near our crew in the cool darkness, bellowing over a hand-held loudspeaker for everyone to get off this field and walk all the way back across campus to the field near the football stadium. The Big Guy walks over to the overhead man and starts yelling in his face, quite rightly, that if his stupid people will get out of here and leave the sheep alone, the sheep will graze quite fine and we will all be able to sleep. As the camp boss sputters to reply, the Big Guy further adds the information that the Hotshot crews on this field have gone with only a few hours sleep in three days and will be getting back up in less than three hours, so unless he wants the fire boss to come out personally, we're not moving.

Vanquished by the Big Guy's aggressiveness as well as the truth of his argument, the camp boss leaves and peace settles back, for the moment, onto our ballfield. The bighorn ram grazes quietly again, although warily, and all of us try to sleep.

A Palm Springs Bash

At three-fifteen I wake to the sound of a jeep whining toward the end of the field where the bighorn is, and sure enough, they're at it again. This time two cowboys are whirling ropes at the dodging ram, and amazingly enough, one of them actually loops him after a few misses. Then it's a wild uproar of spinning bighorn and leaping camp workers, finally ending with a tied-up sheep disappearing quietly off the field in the back of a big pickup truck. By three-thirty all's quiet for the last time, and all the Hotshot crews sink back gratefully into exhausted sleep.

Undisturbed by the smoke and chaos around him,
a firefighter stops to eat.
Author's photo

Shooting flames rip through a patch of pines with a crackling rush. *Courtesy U.S. Forest Service*

A rain of pink fire retardant drifts down onto both fire and firefighters below. *Courtesy Bob Tribble, U.S. Forest Service*

A helicopter scoops water into its bucket to drop onto the fire's hot spots and Hotshots alike. *Courtesy Bob Tribble, U.S. Forest Service*

Smoke chokes even the sun, giving the Hotshot's world a surrealistic quality.
Courtesy U.S. Forest Service

Surrounded by smoke and fire, the author radios to other crew members.
Author's photo

Gone like many other homes in the path of a firestorm, this house, at least, held no human fatalities. *Courtesy Bob Tribble, U.S. Forest Service*

six

Hotshots on Parade

Breakfast tastes wonderful. Well, at least it tastes like food, and at this stage in our hunger-eating cycle a big breakfast is welcome even in the chilly darkness after very little sleep. Breakfast at firecamp is one of those meals you enjoy much more if you're been careful not to watch them fix it.

As they do at many big firecamps, low-risk convicts under the supervision of the California Division of Forestry and their prison guards cook the long-awaited food. Powdered eggs, poured out of a big box into kettles of hot water, slowly turn into bright yellow batches of scrambled eggs. Simultaneously, frozen packets of hash brown potatoes spring to life on huge, portable stoves in a sea of grease beside their accompanying rows of sausages.

For those too queasy at such an early hour to risk the cooked delights, big firecamps, once they're fully established, furnish fresh fruit, dry cereal, bread, milk, and juices, at least as long as they last. When all else fails, there's plenty of coffee.

We try a little of everything, stuffing our stomachs to the limit before boarding a bus that drives us to the nearby heli-base set up just out of town at the bottom of the steep hills. Crew after crew pours out of the steadily arriving buses, lining out gear and men in neat clusters all around the loading area. Then the four helicopters coordinating in this operation begin ferrying crews up across the miles of blackened mountainside toward the southwest end of the fire.

"Hurry up and wait" is the story of our morning, but finally, after hours of waiting, we load into a noisy S-212 and strap our belts tight. Already one helicopter has crashed on this fire, although fortunately no one was seriously injured. Still, the sense of potential danger on every helicopter ride keeps us alert and on edge. Quickly we're off and up, skimming the lower slopes as we speed over blackened desert. The flight seems too short in contrast to the hours of driving it took us to get to the same general section of the fire before, but the direct, straight-as-an-arrow flight of the copter is amazing. We hover, touch down, and an attendant approaches our ship, pointing out a safe line of departure.

All together again, our crew lines out along the ridge behind the Big Guy, following a huge fireline already started by the first Hotshot crews flown in this morning. In all our previous fires we've never worked on one section of line with so many other Hotshot crews. The big bosses have thrown almost 400 men into a major operation on this last, uncontained section of line, changing a previously wild, unpopulated ridgeline into a construction site with swarms of men (and our two women) cutting, chopping, and scraping in a massive assault.

Every few hundred feet we come to another Hotshot crew, passing six, seven . . . ten, eleven, until looking back along the great rounded ridgeline of mountains we can see tiny yellow-shirted figures stretched out for almost two miles, each crew shredding the hilltop of all vegetation in a huge swath. We pass the Redding crew, snake through thickets of brush and small trees for another hundred yards, and then begin, adding our own chainsaws to the already loud chorus of blaring noisemakers. This is truly a fireline of Hotshot proportions—miles long and thirty to fifty feet wide, depending on the fuel type of the vegetation.

The day passes in a blur of cutting, stopping for a crew to move past us, working again, then leapfrogging ahead ourselves. It's a Hotshot parade, hundreds of elite firefighters groveling beneath a baking sun that cooks us mercilessly. But there are no spectators here to watch us on this wilderness ridgeline, just participants—much to the delight of most of the hardcore Hotshot superintendents. They commonly express the sentiment that no one but other Hotshots knows much about the art of wildland firefighting, not even the fire bosses or other big-wig brass back at firecamp who determine all that ultimately happens out here on the fireline. To the Hotshot "supes," only another Hotshot has the ability to judge the performance of any Hotshot crew.

So here we are, on display for each other, each crew strutting its stuff like performing peacocks. The saw teams hold the most visible positions, leading the rest or holding them back by how fast they can slash with the least-wasted energy through the forest of protruding branches and trunks of the brushy thickets ahead of them. Cutting such fireline is almost artistic if done by a real veteran, whose unhurried but effective motions carve out the quickest opening with the least work.

Behind the saws come the toolers, cutting, chopping, and scraping the mineral soil fireline strategically placed in the canopy cut to be most effective. Scattered amongst each twenty-man crew are the superintendent, foreman, and squad bosses. Even on the open skyline of a ridge, often sections of the crews may be working in a low spot or gully where they're totally out of sight from the rest of the crew. By keeping key, experienced men with every group of firefighters, there is always both leadership and caution with each segment of the workforce.

Today most of the Hotshot crews are monitoring the same "crew net" channel on the radio. Although there's always potential for an unexpected fire run, the low-key atmosphere soon gives birth to a lively banter back and forth between crews as wisecracks and humorous insults fly. The joking on the airways turns out to be just about the only thing breaking the monotony of an otherwise long, uneventful shift, as long hours pass slowly as we work toward the final tie-in.

Evening cools the ridgetop, and before we expect it, we've tied in our last remaining section of line. All along the line other crews hurry to scrape the last few hundred yards, and at last the welcome news spreads quickly up and down the line. The Dry Falls fire is now officially encircled, or "contained," as the press releases will state, tied in at last.

High on some unknown grassy knoll just off the sharp ridgeline we camp out a last night under the stars, huddling low in hastily dug "cribs," as we call the rude dirt pits we've chopped out of the rocky hillside. A strong cold wind whips upslope, searching out bare skin and damp clothing on those unfortunate enough to end up with no paper sleeping bag. Hundreds of firefighters spread out in social clusters all over this one-night town. The starlight reveals a great jumbled pile of garbage, cardboard boxes, tins, plastic bags, and assorted odds and ends. Such trash is a sure sign of man's presence in the wildlands.

Tiny pinpointed lights shining from headlamps dot the

landscape on all sides of the knoll. Centered around the biggest campfires cluster the loudest storytellers, each one elaborating in detail on some past fire exploit or phenomenon. And each speaker has a receptive audience, for almost everyone on the mountain feels some addiction, or at least a strong attraction, to the drama and excitement of fighting a fire. Listening to others' thrilling fire experiences only enhances those feelings.

Our enemy, if you can call fire that, is always the same, yet even to the most experienced fireman, often unpredictable. In the same way that a local weatherman often misses his prediction for tomorrow and seldom is accurate in a long-range guess, so fire often defies everyone's expectations.

Perhaps the most intriguing aspect of our challenge whenever we fight fire is the stakes involved in the outcome. By taking those steps away from the safety of our crew buses or helicopter and moving out into the wildlands to face wildfire, we instinctively know we're gambling with our lives. It's a safe gamble, or few of us would even consider it, but seldom does a year pass without one or more firefighters dying in some terrible rush of flames. That element of risk may be the very reason many are here, enjoying the adrenaline thrill of living on the fine line of danger. It's certainly the key ingredient in most of the stories. The adrenaline can be activated by just hearing again of a near escape from a hot run or the drama of racing a "going" fire all the way to the top of a ridge.

All across our darkened knoll, tales are being told, exaggerated, and stretched still again until many over-the-hill firefighters are near speechless, aglow in their not-too-accurate memories of "real firefighting" days. And like aging athletes, many go to sleep deciding to stay on one more season, hoping to experience a few more heart-stoppers, the kind where all the mountains are burning around you for as far as you can see.

But even in the midst of this last-night reverie, reality intrudes in the form of miserably rock-hard ground and biting cold wind. Even young men, not bothered by aching backs or aging knees, still pledge silently that this is absolutely the last year they will ever sleep in crusty, sweat-caked clothing, half-sliding down a rock-hard mountain with gnawing hunger and a cold breeze reaching down the back of their necks.

Hotshots on Parade

The stars sparkle, the crescent moon glimmers, and I'm very, very tired. This fire may be over, but where will the Hotshot parade take me from here?

Fire rages through desert shrubs as wind whips the flames.
Courtesy Frank Podesta

seven

Desert Wildfires

"So you want to be a Hotshot?"

I smile pleasantly as I scan the new faces joining our half of the crew, but I can't keep a little humorous sarcasm out of my voice. There have been many changes in the past four years since my rookie season with this Hotshot crew, three seasons of which I've worked as squad boss for the handtool section of our team. But now I face the prospect of training four new firefighters out of the nine people in our back squad, and two of them are rookies who've never even seen a big fire before now.

"Well, some things I can tell you, some things I can show you, and some things only fire itself can teach you. You're all in top shape, or you wouldn't be here, but after six weeks of running three miles or more every morning and then working hard all day, you'll feel even fitter."

I nod toward Milo, Nell, Jack, and Dave. "Watch and listen to them. Never, never, never stray away from your squad in a hot fire situation and don't be afraid to ask questions. Just don't always expect a detailed explanation in the midst of the action." Scanning their faces, I see a mixture of bravado and nervousness, both to be expected from new crewmen. "Remember, a Hotshot crew goes where no other crew will go and works longer and better than any other crew will work. If those prospects scare you, drop out now, because once we hit the fireline we'll be depending on you to pull your weight. Perhaps our lives (I nod toward our experienced crewmembers) will rest in your hands. At the least, we don't want to do your work. Okay?"

My words haunt me the next night as we wind down over the far side of the rugged crest of the Sierra Nevada range toward our first fire of this season in the desert beyond. Will these untrained crewmen endanger others, as well as themselves, if this fire is a "gobbler"? How can I teach them quickly the safety awareness and caution that comes second-nature to most experienced firefighters?

Jack shifts restlessly against the window as he unsuccessfully tries to find some comfortable position in the cramped quarters of the front seat of our crew bus. His frustration also reminds me how tired we'll all be when we reach the fire after traveling all night. I can already feel my eyelids sinking, my drowsy concentration drifting . . . drifting . . . only to jerk up, heart pounding and frightened. I pull back into the middle of my lane. Was I asleep for more than a second? Shaken, I glance in the rearview mirror back through the open boot of the cab at the seven slumped forms twisted on small hard seats. No one's staring back at me with terror. I must have just swerved slightly.

Jack shifts position again, then sits up sighing in frustration. "You want me to spell you?" he finally asks, but his giant yawn and swollen, bloodshot eyes remind me how drunk he was when our late-night fire call came. Knowing he hasn't slept, I trust him even less than my own sleepy self.

Fire assignments come at random during fire season, often at night and often on our days off. Over the years my wife has learned the hard way not to count on me for any important occasion during fire season. In fact, whenever fire season is slow and the big money-making fires just aren't burning, she jokingly suggests planning a dinner engagement and movie, complete with a babysitter for the kids. Past experience has taught her that any plans during the summer or early autumn always fall victim to the ring of the telephone or the squelch of my pager.

Nighttime phone calls are truly classic, straight out of some old-time comedy hour. We'll be in bed, just falling asleep, at eleven o'clock or so, when the phone rings and I stumble through the darkened house to get it.

Paul or the Big Guy will be there, expectant excitement in their voice. "Call your people, we've got an off-forest assignment! Be at the warehouse at four in the morning. We'll meet the plane at six."

If I'm halfway awake I manage to ask where we're heading, but often this simple question slips my groggy mind and I begin the thankless task of trying to round up seven or eight firefighters over the telephone. It really never matters where we're going, whether it's Washington or Arizona, North Carolina or Utah, there'll always be a big fire at the end of the journey.

Ironically, the last message Paul or the Big Guy always gives me is "Tell everyone to get some sleep . . . we won't be getting any for a while." As every Hotshot knows, traveling is easier with some minor preparations, so even though we're always expected to be fully prepared, there are still questions like "Did I pack my extra socks?" or "Are all my t-shirts stuffed in the dirty clothes basket?" that run through your head.

So the phone calls begin, and by midnight, everyone's usually accounted for except Jack and Dave, who won't return from the bars until two A.M. Their houses have the message, though, leaving our half of the crew complete, information which I call and relay on to Paul. After a quick check of my gear, I slip back into bed, hoping to catch a little sleep before rising at three o'clock.

Sleep doesn't come easily, though, for the sudden anticipation of a faraway firefighting assignment is exciting, even when it happens many times each summer. There are always the mental preparations — remembering which appointments I can't make, whose dog may need feeding, or if I have sufficient money for expenses. There's also the gradual preparation of nerves, slowly building as the time to face a dangerous wildfire draws closer every moment. Simply going back to sleep with casual indifference doesn't happen. There's just too much to think about.

By one-thirty or so I'm falling back asleep, just about the time that Jack or Dave call for any details their roommates may not have heard. I yawn and tell them to get some rest, so by two I'm back in bed and falling asleep for the third time of the night. An hour later the alarm rings. I quietly rise and try to dress, finish packing, eat something, and leave without waking the rest of the family. We trickle into the warehouse one by one, occasionally driven by a wife or girlfriend who drives away with a yawn and a blank look of disbelief that she's actually out at four in the morning.

Minutes later, the Big Guy calls us all together and mentions how lucky we were to at least all get a good night's sleep. No one says anything — there's nothing to say when your mouth is stretched to the limit in a giant yawn.

Almost as frequent as the all-night anticipation dispatches are the midnight rolls like this one over the Sierra Nevada crest to this desert fire near Reno. By hurrying, we'll make it to the fireline shortly after sunrise, hopefully in time to catch the flames before the increasing heat of the day adds to the fire's intensity. A lucky few can honestly sleep on the twisting, curving drive, but for most of us, an uncomfortable, sleepless drive is part of the expected price to pay for the "glory" of traveling as a bona fide Hotshot firefighter.

Just as dawn spreads its golden glow across the multi-hued hills outside of Reno, I catch myself falling asleep for the last time. Too tired to even worry about Jack's condition, I nudge him awake and pull over for a quick change of drivers.

Jack speeds off in hot pursuit of the other bus while I crumple up against the door, using a sweatshirt for a pillow, and relax. Jack's choice of loud music on the radio arouses some initial gripes from those trying to sleep behind us, but all complaints fade when they see just how tired Jack looks. They know Jack and I are the only ones licensed to drive this bus.

We follow the other trucks onto a wide, straight-as-an-arrow desert road that shoots for miles across gullies and over low ridges without an end in sight. I'm fading fast, despite the bounding bumps and shaking from the rough ride when I happen to glance up, yawning, to see us heading straight for a big ditch alongside the road. "Jack!" I yell and we grab the wheel at the same time and one of us pulls it back onto the road. Jack slows us to a sliding stop, looks at me with relief and says, "That's all for me!" shaking his head in either disgust or wonder.

So, awake now, I retake the driver's seat and drive with radio blaring and window wide open toward our fire destination.

"Anticlimactic" is an understatement. The fire, or in this case, both fires, are out. I don't mean smouldering, smoking, or just creeping around. I mean out. The larger fire is more than 1,200 acres, all of which burned during the evening and nighttime hours, and yet there's no visible smoke anywhere. Wait, there's a tiny column drifting slowly up from a canyon near the upper edge of the fire.

Desert Wildfires

We gear up, line out, and wait while Paul and the Big Guy get our assignment from a Bureau of Land Management (BLM) boss. "Cold trail and mop-up!" comes the order, so we divide up into two crews, fan out, and begin to move along the fire's perimeter.

All fires thrive on the wind, but without strong breezes most desert fires could never burn at all. This great arid valley is a perfect example, for its brittle sagebrush and dry grasses all grow in scattered clumps isolated from other clumps two to ten feet away. Only a brisk wind could somehow sweep flames low to the ground, whipping long tendrils of fire horizontally from clump to clump.

Now that there's no wind, we could throw matches into clumps of dry grasses all morning and get nothing more than a flashy flare-up that would most likely burn out without even spreading to the closest fuels near it. To tediously and carefully turn over each scorched clump of wood or dig up each whitened pile of still-warm ashes seems a waste of time, especially to our newer crewmen.

Moans and groans gradually turn into irritated sarcasm and bitter complaints as the hours pass and we still move steadily along the fire's edge. I take the outer ten feet, Nell the next ten, and each firefighter farther in takes his own ten-foot swath, allowing us to sweep the entire area most likely to harbor lingering coals that just might reignite under late afternoon winds. By lunchtime, great cumulus clouds begin to build in the west, each slightly larger and taller as the Sierra Nevada range turns moisture into thunderheads. While we stop to eat, we watch whole sections of sky darken until far-off rumbles reverberate back and forth down the great desert valley.

But no rain reaches us yet, and despite the scattered thunderheads, the scorching sun and high humidity leave most of our crew glassy-eyed and drained. The lack of intense action, as well as lack of sleep, join with the heat to stop even minimal work. We each melt quietly, without planned discussion, into the partial shade of a scattered grove of unburned junipers flanking one side of a rocky draw. There's no soft place to recline, no real shade to cool us, but just enough of a hiding place to escape the worst of the sun's burning hot zenith.

Suddenly radios blare, "Everyone back to the trucks—they've got new fires south of here!" Groggily, everyone lines out, brushing off the tell-tale dust from our fire clothes as we begin the long walk back to our trucks.

Almost an hour later, drenched in sweat and feet aching from trudging through the hot sand, we're finally all back and our gear's stowed away. As always, the stimulation of heading to a new fire perks up all but the sleepiest crewmen, so questions and conjectures fill the new animated conversations. We drive at a fast pace, staying at least a quarter mile apart on the sandy dirt road so that the dust clouds can subside enough for us to see the road.

The flat sameness of our desert landscape gradually changes, mixing with upthrust rock ridges to form broken foothills, and all of a sudden we're back to the main highway and a waiting cluster of BLM pickups and law enforcement vehicles. At the moment we pull off the road to find a place to park, the overhead bosses reach some decision, scattering to their trucks and filing out into a line headed back toward Sparks and Reno.

We fall in behind, enjoying the cool breeze and lack of dust as we hurry along the main road. Much sooner than we expect, we round a curve and emerge into a valley where three fires burn in the distance, two of them going pretty good as evidenced by the huge patterns of blackened hills trailing behind their flame fronts.

Surprisingly, we divert from the highway short of the closest fire to pull into a side road bordering a county fire station. Big fire trucks and fancy pumpers stand useless, for even from here we can see that the two "going" fires are not accessible by road. Uniformed firemen and volunteers huddle in groups to talk excitedly, but talk is all they can do, not being equipped to battle wildfires in such rugged terrain.

A strong wind, gusting to fifteen or twenty miles per hour, blows a fine grit in our truck windows as we wait for the Big Guy to get our orders once again. Paul finally tells us to get out, water up, and sharpen our tools, and none too soon, for here come the television camera crews, eager to film our grimy faces.

We're all sharpening our tools in our most picturesque positions when the Big Guy returns with new orders. "Load up!" he shouts. "We're going to a fire!"

Still undiscovered as future Hollywood heroes, we leave the unpacking camera crews to feature the nicely uniformed county firefighters while we hurry off to the fire. At first it appears we're headed toward the biggest fire, which is burning sideways across the fairly flat hills two miles or so southeast of us, but in the bright light of the lowering afternoon sun, the shapes of a pair of big bulldozers appear on the hottest flank. We're not needed there.

We reach another intersection, whip around a couple of sharp corners, and there's our fire—smoky but not too intimidating, blowing downslope off the crown of a rocky, grass-covered hill. One BLM engine crew seems to be holding their corner of the fire's advance with a steady stream of misty spray, but the rest of the fire's moving right along.

With no one but the BLM firefighters to witness our prowess, we leap out, tool up, and line out quite nicely and quickly, then march determinedly in a tight line right up through the smoke to angle around the hill and approach the fire from the west. Since there are no trees to cut with chainsaws, we all carry hand-tools, so we have an abundance of flailing, chopping bodies to meet the fire's advance. Even so, for a few minutes I fear that the wind is just too much, for it takes lots of time to rip out the dry grasses and desert bushes that somehow survive in the rocky covering of this hillside.

Where the flames are worst, I leave my tail-end position and pull Jack and Nell forward to help me frantically search for sand and then fling it at the flaring grasses before they breach our line. Even the primitive stomping of feet and drenching our shaky line with canteen water almost fail, but suddenly the fire's stopped, smouldering like a frustrated bull at the end of its tether.

From then on it's all mop-up, line improvement, and "take-a-breather" time. We spread out for a half mile along the fire perimeter, holding and widening the fireline, but despite still-strong breezes and brittle fuels, this fire is history.

An hour later, still looking half wild and more than half blackened, our crew carefully files through the doors of the Nugget Casino for a well-earned reward of restaurant food. There are many tales of hunger and just plain lousy meals whenever firefighters meet, but BLM in the western Nevada region some-how understands what we really need. The food at the Nugget is top-rate, the same as it's been in previous years when desert fires outside of town brought our crew to the action.

We're feeling jet-set giddy, twenty of us dominating the restaurant, answering earnest questions from admiring guests, accepting thanks not too haughtily from young waitresses who've watched the fires on the news.

It's steak and seafood, chocolate shakes and french fries. We're ready to go on for days, totally forgetting our missed night's sleep and strenuous day. Supper's end, near midnight, brings a still-excited crew together in the brightly lit parking lot to joke and slowly recollect the day's highlights.

Then off to a short night's sleep on a college's damp green lawns before we'll rise in the morning for our whispered next assignment — defending the infamous brothels south of town. Although our two women chide the loudest braggers, overall the crew is already intent on a special show of excellence for tomorrow's hoped-for audience. I kid Jack that it won't matter either way, the ladies will all be resting during the day, but logic can't stem the steady stream of jokes. Eyes sagging, one by one, firefighters fade into their bags, and tomorrow's possibilities turn to sleepy mumbles and yawning nods of bleary-eyed exhaustion.

Even in such dry, rocky terrain, the desert fuels burn fiercely, putting out oily, black smoke.
Courtesy Frank Podesta

eight

Cat Houses to Joshua Trees

A restaurant breakfast and a warm, sunshiny morning combine to invigorate our whole crew. Not until we've arrived at the large-but-sputtering fire itself and started the long hike up the sage and dry grass-covered slopes do the first signs of tender muscles and bruises begin to remind us what the summer holds in store for all of us.

Without objects to compare their size with, most desert mountains don't seem too large . . . until you start climbing. We finally pick up the erratic fire's edge halfway up the slope, just as the first morning breezes add enough oxygen to create some minor hot spots right ahead of us. Amazingly, grasses and sagebrush are growing in a hillside of nothing but broken, crumbling rock, so no matter where you chop or how hard you try to find sand to fling at the flames, all you get is a metallic ringing for your reward.

Like our first fires, only the breeze keeps this fire active; in fact, ninety percent of the fire's edge is already out from the lack of wind during the night. Steadily, but not too successfully, we flail away at the sputtering, leaping, diminishing, leaping-again flames, gradually picking up most of the remaining hot spots when our radios announce the arrival of the first retardant drop of the day.

Not really needing any help on our section of line, we're puzzled when an S-2 drop plane roars low right above us and on over the ridge. Hearing that a drop's coming in on us anyway, half the crew moves into the cool blackened burn while the rest of us scramble downslope. Seconds later, the S-2 rumbles out of nowhere, whips past in a thunderous pass, and then the hiss of raining red retardant descends in a brief but drenching shower.

The drop lands right on the money, not only extinguishing

every remaining flame along this whole hot section, but also turning all the Hotshots who moved into the burn a bright red. Shrieks of indignation and screamed curses gradually fade as we work the fire's edge on up the mountain, climbing for hours over endless boulders and patches of rocky grit until we've cold-trailed the whole western half of the fire.

The high point of the morning is the appearance of a lone photographer who's somehow scrambled all the way up the mountain to find our crew. Lacking any good action shots, he's delighted when Ken gets a good burnout assignment, lighting off an unburned pocket of grasses and sage near a rocky outcropping. The following day's paper will feature a color shot of Ken and the flames spread right across the front page of the Reno paper.

The work goes slowly, as a stronger wind whips up after the photographer leaves and steep terrain prevents easy access to patches of burnt brush amidst sliding rocks. Far down below, scattered trailers and a couple of houses may or may not contain the rumored audience of nighttime ladies that our loudest, most macho guys had hoped to impress. As usual, a few red fire engines sit parked near the bottom of the hill, their uniformed crews appearing only as tiny dots as they cluster together to drink coffee and watch our progress far above.

We break for a short lunch, then discover a half-mile section of white-hot embers, each little cluster the still-smouldering base or roots of a large juniper, pinyon pine, or sagebrush. For hours we painstakingly dig up these embers all along the dead edge of the fire, mixing them with sand and crushed rock, chopping the embers into bits and spreading them to burn themselves out while we're still there.

Every hour the winds grow stronger and gustier, blasting those of us farthest up the ridgeline until raw skin begins to crack from the steady sandblasting grit. It's the "mindless hour," the time to block all attempts at rational thoughts from your mind and to simply grovel, methodically digging, bending, walking, until the afternoon winds slowly fade and at last we're all tied in to a completed fireline.

No welcoming reception awaits us at the bottom of the mountain; in fact, everyone's gone except for a couple of patroling observers and the Big Guy. Just as we load into our buses, one of our rookies starts throwing up all over our rear tire, not the greatest send-off for our drive in to dinner. His head is forced out the farthest window back, and oblivious to his misery, the buses roll out as we lean back in the comfort of real seats after a long day of endless walking.

Three big fires in two days—five days into the season. We're on a roll, following the wind-driven flames. Already the latest rumor's filtering through the crew. Halfway back toward Carson City it finally reaches Jack and me in the front seat. We're headed to Death Valley!

The next day, in typical off-again-on-again fashion, we're going, we're staying, we're going, staying, until finally we're really going. The long, monotonous, incredibly hot, windy drive ends in the still-warm darkness of midnight amidst a cluster of trucks, buses, and equipment. Everything is silhouetted by a giant pale moon that casts its spotlight glow far out over the mountains of Death Valley National Monument.

It's bed-down time in the sand and dry bushes alongside the sandy dirt road we followed for miles into the fire. I turn and half-jestingly warn the four newest members of our bus, "Some of you might not have slept on the ground in the desert before, so don't let me worry you, but there are a couple things to remember.

"First, there are rattlesnakes everywhere, but it's unlikely any will actually crawl into your bag, despite the fact that they may be attracted to your body's warmth. Then there are tarantulas, gila monsters, and two kinds of scorpions. The tarantulas won't kill you, but their bite is supposed to be extremely painful. The gila monsters, although very poisonous, are not aggressive unless bothered, and the smaller of the scorpions isn't much of a problem.

"But I'm sure you'll want to know that the big orange scorpion is extremely deadly, so watch out for it. Anyway," I continue, "have a good sleep, cause we'll be getting up at our usual early time."

Without much further conversation, all the experienced

Hotshots unroll their bags, flop down, and proceed to the serious business of sleeping. I notice as I lock the truck, though, that all four of our new crewmembers seem particularly nervous about where to lay their bags. Even as I snuggle a hip-hole in the sand, I can still hear rustling and low whispers as they survey the countless holes in the dry sands amidst the broken boulders and piles of dead brush. Far off the yips of a lone coyote linger on the breeze.

Despite official predictions of 115° temperatures and unbearable conditions, the Death Valley fire turns out to be mostly a low-key mop-up shift. After only one day and night, we leave behind a big fire that's totally cold to head south again, this time to Joshua Tree National Monument where yet another gobbler's on the loose.

One of the true privileges of being a Hotshot is the chance to see the remote, unusual natural wonders of America. Given the right conditions, almost every type of vegetation can burn, so within a single season we may visit high, cold forests in Idaho, desert fires in Arizona, range fires in Nevada, leaf-littered hardwood forest fires in Georgia, and brush fires in Southern California.

Joshua Tree Monument is one of those exotic places that seem unreal even when you're right in the midst of the scene. We arrive in sweltering heat and clear skies, wondering just where this 3,000-acre fire's supposed to be. We wind in along the monument's main road, entering a large valley covered with cacti, low bushes, and great numbers of the biggest Joshua trees I've ever seen, until finally we reach a hastily assembled heli-base.

We wait for almost an hour, taking shelter from the hot, gusty winds alongside of familiar heli-tac crewmen we know from years of previous fires. We still can't see much but a little smoke from the fire that's three or four miles distant across the cactus-studded valley. No roads go anywhere near the fire area, so it's all a fly-in show. We hurry up and wait some more until suddenly in a rush of activity we're flying out in squads of six or seven, landing in a sandy flat at the very base of the mountain.

Even before we land we can feel the vicious gusts blasting our copter, slamming into the ship with enough force to knock us sideways in the air. As we land roughly and hurry away from the ship, bright-orange flames flare up all along the hillside above us, driven down at us by the wind.

I pull together my squad, share three toolers with the saw squad, then head south with the rest to head off a whole section of flame a quarter mile away. As we half-hike, half-run along the sandy, brush-covered flat at the base of the rocky slope, I can see only too well our problem. We'll have to try to stop the fire right here at the base of the hill, for the bone-dry, tall mesquite bushes and crowded groves of taller-still Joshua trees that exotically cover the flat valley floor provide such a consistent, intense fuel that we'll never be able to slow it if it gets out into the flats with these winds on it.

Glen, a veteran with more than ten years of Hotshot firefighting experience, has accompanied us to fill in for a missing crewman. With one helper, he begins a frenzied assault on the closest flare-up while I direct the others in an equally crazed attack on the other two main fire heads. With the wind whipping the flames horizontal to the ground, each cluster of brittle bushes or dry grasses almost explodes with a hopscotch of leaping fire.

Nothing's effective but sandy dirt flung by the shovels, which we use in a flailing, scooping flurry as fast as we can dig and throw. Most of the crew carries the Pulaski, good tools elsewhere but nearly useless for throwing sand to smother such intense flames.

I knock down one hot spot, rush to the next, only to watch helplessly as two small heads of flame veer apart and push right out into the flats. Caught up in my own hot spot, I yell futilely for assistance, already aware that everyone close enough to help is already beating unsuccessfully at their own hot spots. My attention falters, half caught by the other two hot spots, and my own hot spot flares up into the bushes I've been fighting to keep it out of. Heat sears across my cheek, flames blast past my right ear, and I stagger, stumble over a bush and leap back up to begin again.

The heat's intense, but Dave and Milo rush to help, having knocked down the hot spot they were working. Just as we suddenly halt the leaping flames with a coordinated assault of sand-flinging, chopping, and stomping, I notice Glen, standing between the other two hot spots, single-handedly pouring such a stream of sand onto first one, then back at the other, that he holds both flare-ups long enough for us to aid him.

For thirty minutes or so, it's an all-out battle played out with no spectators but the endless desert and the rugged, rock-strewn mountain that rises above us. We win this skirmish, thanks mostly to Glen's incredible efforts, which leave him with a broken handle and mangled shovel, but the victory's taken a lot out of our crew. Yet there's no rest now; already the Big Guy is on the radio, calling for reinforcements to pick up the northern flank as it pushes its way down off the mountain north of where we landed.

For the next few hours we leapfrog along the sandy wash that hugs the foot of the slope, burning out the fuels ahead of the fire whenever possible and chasing the flames out into the flats whenever long fingers of fire reach down off the fiery hillside in quick thrusts that always seem unexpected. By dusk we're feeling confident, hopeful we've finally got a handle on this, the hottest side of the fire. I look out over my crew, most of them hungry and beat and none of them enjoying the spectacular desert sunset colors that turn the grays, greens, browns, and whites of this Joshua treed valley into their own fiery hues. How long can we keep this pace up?

We turn a corner around a finger ridge, only to hear radios blaring out calls for help from up ahead. Almost the entire crew surges toward the spot fire, already two acres in size and growing every moment. Like our first hot spots, flame-thrower blasts of horizontal fire shoot from bush to cactus to bunch grass with unstoppable intensity until blasted into smouldering quiet by a combined assault of many well-thrown shovelfuls of sand.

We circle the spot fire from both sides, losing the head of it, then squeezing, then squeezing in some more. At last, fifteen of us all work together to smother the last flames. Then we flop down for a needed breather.

Almost a mile ahead, the Big Guy's calling still again on his radio, urging on the burning squad to regroup and continue firing while the rest of us clean up the spot fire. Far into the darkness, we battle hot spots and windblown flames around the curving shoulder of the mountain, until long after midnight, we finally carry the black fire edge into a solid rock drainage that ties in with a whole cliff of sheer rock bluffs.

Unsure of what's ahead in the darkness and sure that we couldn't care less anyway, we spread back out over the last mile or two we've held, checking for spot fires and taking a long overdue lunch break.

We sit in groups of two or three, each group a good distance apart, each silhouetted by their little warming fire as we huddle around the radiant heat of the flames to regain the day's lost warmth. The temperatures have plunged dramatically, falling most in the hour after sunset. Wringing sweat-drenched clothes out in front of the fire, it's times like these that somehow finalize the addictive side of firefighting. We congratulate each other for minor heroics, agreeing wholeheartedly how great we all were, how much we've saved by our efforts, and how incredible the backfire firestorm looked as it gobbled up the entire northern slope of the mountain.

As sick as we are of fire, it's still the main topic in each little group, with almost everyone talking about some odd phenomenon of fire behavior or strange quirk of the fire's advance.

Fire appears to be primitive, simple, and easily understood. It's our opponent, but here we are huddled around our own little fires, absorbing each fire's precious warmth and light in the chilly night breeze—so fire is almost our friend. And although the devastation of a wildfire can be easily understood, there are so many sides of fire that are so complex, so unfathomable, that even the best computers can only approximate general predictions of how it will behave.

Tonight we talk about fire as if it were an entity, alive and able to choose, for often it seems that there's no explanation for why the flames pass over whole sections of bone-dry grasses, only to veer away into a slope that seemed nearly void of fuels. We sit around our tame little fires, feeling like conquerors, knowing only too well how futile our efforts will be in the days and weeks ahead in trying to stop the blowing-going firestorms of a truly out-of-control wildfire.

To regular firefighters, who work only in one region, such awesome, overpowering fires may exist only in stories, especially when only one such fire may burn in their region in a whole decade. But to Hotshots who may see three or four heart-stopping, thundering blowups in just a single season, fire becomes a powerful high, a thrill that sends adrenaline racing and brings the senses to their fullest awareness. The knowledge that our lives depend entirely on just how quickly we react, how carefully we choose our routes, how safely we take our stands, all forces us to be totally alive, right now, in the moment, ready to run, hike, take cover, or fight fire as the need demands.

So now we sit and talk, reliving the natural high of the danger and excitement that's filled our last twelve hours. Although exhausted, we stir to look out for a last time over the hundreds, the thousands of glowing mesquite and Joshua tree stumps, flickering like grounded stars. I walk back up the line, offering band-aids and moleskin for cuts and blisters, offering humor to those in good spirits and warnings to those vocally critical of all the day's tactics. With such a large crew, nothing ever pleases all twenty people, so there's sure to be one or two critics unhappy and quick to broadcast their anger.

Overall, the danger and yet the success of this suppression effort satisfies almost everyone. I pass on the latest official word: our crew single-handedly stopped the entire burning edge of a 3,700-acre fire. We'll all enjoy repeating that one at the next encampment of Hotshots on Parade. But the glow of success can't ease the discomfort of the rocky sand and our far-from-full stomachs. As dawn's first pale light spreads across the eastern night, I'm fighting off exhaustion and a multitude of new aches and pains. Ah, the glory of the Hotshot life. It'll make a better tale than it feels to live it tonight.

Mop-up in heavy forest often takes many days of constant turning and stirring.
Courtesy U.S. Forest Service

nine

Up to Idaho

"All right! Idaho!" echoes from at least five mouths at the same time. And probably all twenty of us feel the same way. Even after two months of firefighting and project work, the call to go to a different state adds excitement, not to mention some sure overtime, to our assignment.

After two hours of driving to Sacramento's airport, we're airborne once again with the Lassen crew, headed to Boise, Idaho. Although we're not consistently paired with any one crew, it seems we've often ended up traveling with one particular crew over and over again. So we've got lots to talk about on our flight — fire talk to catch up on and friendships to renew. Just before daybreak we file neatly across the Boise airport runway, sprawl on a small lawn outside some offices, and as usual, try to get some needed sleep.

Our day is filled with bus rides, transfers, then more bus rides, until midafternoon finds us swaying in our bus seats, winding up a narrow road into the high mountain forest of central Idaho. Firecamp is really beautiful, at least from a scenic standpoint — a five-acre meadow surrounded by nice conifer forest just across the road from a marsh with beaver ponds and streams. But an overabundance of fire overhead bustles about camp, snapping orders, motioning directions with curt tenseness to truck drivers and crews. The whole attitude at this camp is different — not the urgent but comfortable "here-we-are-again" attitude that fire-prone regions like California generate, but more a bristling show of tough appearances by fire overhead personnel who seldom handle big timber fires and perhaps feel unsure and tentative.

For whatever reasons, tempers are short, the food's greasy, and it's a relief when we're bused for the last time, this time back out of camp to a spur road and trailhead closer to the fire. It's evening by the time final directions are agreed upon and both of our Hotshot crews are strung out and geared up, waiting. Then we're tramping through the woods, contouring over ridges, past the last stumps, and on up into the wilds of a rugged but beautiful Idaho mountain range.

We run out of trails, bushwhack through tangles of bushes and thickets of young trees for at least another hour, then delicately scamper across a nice, shallow little stream before the smell of smoke alerts us to the fire. Even then, the fire seems just about out; the only sections we can see are just creeping around in alder and aspen thickets or sputtering about in the dry pine needles covering the understory of the open pine forest up from the drainages.

We anchor at the creek, cluster for a few minuites while the Big Guy and the Lassen superintendent agree to split up the two miles of open line between our two crews, then we're off and humping, sawing and pounding the ground in not-too-successful efforts to find mineral soil that won't burn beneath the foot-deep layer of humus and matted roots. The need to cut fireline down to mineral soil varies from fire to fire, for along damp, riparian forest floors or in the wet-type climate zones of a coastal forest, just pulling back the dry leaf litter or dead branches may be enough to stop the fire's spread. Often, though, in the deep, dense layers of a conifer or hardwood forest, days after the surface fire's been totally mopped-up, smouldering ground fires can still creep inch by inch through the humus, gradually spreading until they return to the surface outside our fireline and a brand-new fire ignites, taking off again. In Alaska firefighters must deal with frozen tundra or bone-dry peat-like fuels, depending on the time of fire season. Once in a virgin, old-growth forest of ancient beauty and equally ancient forest humus, we dug down more than two feet in many sections of our line before finally reaching soil not intermixed with dry forest litter.

Here on the Idaho fire, it's a long night's work, the kind that takes a little extra, especially from the tooler, struggling through the dense ground litter. Firefighting for a hand crew is a team effort, but often the work required by part of the crew doesn't come close to matching what's required by the rest.

Tonight, the front half of the crew, the saw teams of sawyers and their pullers, have an easy time. While our tool squad is slashing and scraping at high speed, taking only a few minutes' break every hour or two, we repeatedly find the saw teams flat on their backs, resting comfortably as they watch us cover only one or two hundred yards in a steady hour's work. By midnight, our squad's really getting ragged, stumbling over roots, missing on the slashing chops of protruding brush or small trees, and yawningly wishing they'd managed to sleep more on the flight last night.

I scout up ahead, checking to see how much fireline we have left to build to reach the proposed tie-in point. Far upslope, beyond thickets of brush and more aspens, I spot a bright glow, and beneath it I see the jubilant front squad, brewing coffee and hot chocolate, laid back around a big warming fire, laughing and having a good old time in a large, pine-needle-covered hollow. Our half of the crew's still a quarter mile back, drenched in sweat and just about spent. Somehow the balance isn't working. I move up and explain the situation to Paul, who listens to my progress report, discusses it with the Big Guy, and sends me back with the message that once we reach here, we'll stop till dawn.

On my return, I downplay the comfort and lack of support from the rest of our crew, instead emphasizing that the work's over for the night once we reach the hollow. Even that task proves difficult; for the next two hours cutting and scraping turn out to be even harder than earlier. Big fir snags, upslope from our digging, crumble and fall twice in fiery, crashing eruptions that both times scare the living daylights out of those of us working down below. Neither time do the embers or chunks of flying, burning branches actually cross our fireline, but the shock of their falling shakes the ground.

By three o'clock we're finally nearing the long-awaited hollow, desperate for a few hours of shut-eye. I tie in, then each of us works back toward the rest, hoping to show the lazier half of our crew what we wish they'd done earlier. Just before four o'clock we're done, tied in to a wet drainage right at the hollow's edge.

We scatter out amongst the rest of the crew, wrapping ourselves in the noisy rustling of our aluminum foiled space blankets to try to hold in at least a little body heat in this high mountain

Idaho wilderness. My eyes close, I'm so happy just to lie still, when just at that moment I hear the Big Guy called on the radio. The crew needs our help on up the line.

The Big Guy knows we deserve the rest, but firefighting policy frowns on fireline resting. He can't really refuse to help, so after stalling for a couple of minutes, he passes the dreaded order. "Get up! Let's go!"

Our squad almost simultaneously turns to look at me with disbelief, then betrayal. I start to catch up to the Big Guy to remind him how hard we've labored to earn this rest, but sighing, I stop. There's no way to change things now. "Sorry, let's go!" I urge, and we struggle up the mountain to help the other crew work the remaining line.

The remainder of the night and early morning are a blur, climaxing when Milo gets the worst of a meatbee encounter that ends with a sting right on his nose. Despite everyone's heartless jokes, Milo handles the pain well, but soon his nose is so swollen it looks inflated. We cluster around him raggedly, trying to bolster his spirits with the best that humor can offer until scattered back to work by the Big Guy's anger.

The new day's sunshine brings not only a completed fireline but a warm breakfast flown in by helicopter and some up-to-the-minute gossip. According to camp rumor, this fire may or may not be related to a recent conflict between a bear and some sheepherders. Apparently the bear suddenly appeared one night, rushed the sheep, panicked the whole flock, and all 200 or 300 of them ran headlong right over a cliff to land far below in a terrible mangled heap right smack in the middle of a nice-sized stream that borders one flank of the fire.

All of this supposedly took place a week ago, but since then, someone set traps in this pile of rotting sheep and managed to catch and kill one or two black bears, depending on who tells the story. The total resulting mound of decay has contaminated the stream as well as released a quite distinct aroma along that whole flank of the fire.

Whether this fire has any connection, no one knows, but since there's been no lightning, arson appears a likely cause. After hearing the story, we all agree unanimously *not* to sample any water from any stream in the entire area, at least until we're sure which one has the sheep. In fact, for the remainder of our time here, it's amazing how popular canned juices become, even among those who never drank them much before this.

We hear this same bear story from enough different sources that it appears to be fact, adding another chapter to our continually growing saga of conflicts relating to wilderness. Should livestock be grazed in a wilderness? After walking in for miles from the end of the nearest dirt road, it's always a letdown to find herds of cattle and their prolific droppings scattered all across a wilderness valley or wilderness stream. Or should mining be allowed in wilderness? On some fires we've seen freshly bulldozed roads carved across some wilderness hillside so that some small mining operator can easily access his mining claim. Is such damage a good political trade-off, or natural damage we shouldn't allow in a wilderness?

Questions like these arise often when we fight wilderness fires, for suppression efforts may take us away from the main trails and most-frequented camping spots to the more remote corners of these wilderness preserves. And while loggers, ranchers, and developers generally have little use for wilderness preservation of any sort, even environmentalists sometimes disagree on exactly how to manage these last remaining wildlands.

The questions of how to fight fires in such wilderness areas brings just another controversy to wilderness managers. Fires are nothing new in America's wildernesses. On the contrary, thousands of years of lightning-caused wildfires have created much of the fire-adapted ecosystems that cover the deserts, mountains, and prairies of the American West. Yet man has relentlessly battled fires in wild places, especially since developing the technology and equipment to make these efforts almost routine.

Part of the reason for fighting wildfires lies in their potential to spread, escaping wilderness boundaries and threatening commercial stands of timber or recreational areas or even communities. But the politics of firefighting are so complex that often wildfires that couldn't possibly escape the rocky perimeters of some high crest-zone wilderness are still aggressively attacked for reasons not quite clear to the firefighters themselves.

An American institutions is to "do something!" when faced with a questionable situation. The same philosophy, applied to wilderness fires, often destroys more of the true wilderness than any fire damage could ever do.

Weeks after the Idaho fire, we're cutting fireline on a lightning fire in a remote corner of the Trinity Alps Wilderness in Northern California. Located deep in a densely forested bowl of a valley, the fire made a good run the first day, almost died out during the night, then rekindled enough the following morning to thunder right past our exhausted crew and devour a strip of forest right up and over the ridgeline.

A stream of Hotshot and regular forest crews arrive to aid us, each flown in by helicopter in a massive airlift involving hundreds of people, tons of food and supplies, and great amounts of firefighting equipment. Although the fire itself appears burnt out, with only minor puffs of light smoke occasionally drifting up from the valley forest far below, the combined forces all begin cutting a major fireline along the rim around the entire valley.

Day after day, even during light showers and cold, foggy weather, our crews slash a twenty- to thirty-foot-wide "highway" through this wilderness, cutting hundreds of large trees and stripping the middle ten feet of the line to bare earth. Finally, when conditions are favorable, helicopters dripping napalm ignite the whole valley, thousands of acres, in a huge, fiery burnout operation that multiplies many times the damage the fire might have caused if left alone.

Final cost? More than a million dollars and a gigantic scar on the wilderness — not from the fire, for even a major fire will look natural and provide wilderness habitat within a few years or a decade at the longest. No, the scar left is the "highway" fireline of cut forest and denuded ridgetop — impacts that will be visible for far longer than the effects of the fire.

Was the fire really a threat? Did it need such a massive suppression effort? Wilderness firefighting often brings such questions to the Forest Service, Bureau of Land Management, and Park Service officials who make the final decisions. Politically, it's much safer to err on the side of aggressive firefighting, for who can fault an official for "protecting the public safety"?

To err on the other side, however, is politically disastrous. Imagine whose head is on the block when a "safe" wilderness fire somehow escapes and ends up costing millions of dollars in damage, or even worse, a human life.

Even though the major decisions are made far away in some supervisor's office, it's often the bosses of the crews on the fire that ultimately determine just how severely to chew up the natural features of a wilderness. One example of this control occurred on a 4,000-acre lightning fire that had been left to burn in Yosemite's southeastern mountains. Two crews, one from the Park Service and the other a Forest Service Hotshot crew, were finally assigned to stop its spread so its smoke would stop blowing down into Yosemite Valley where it choked employees and tourists alike with its particularly dense, stinging effects.

The park crew went one way, the Hotshots the other, and by nightfall a fireline encircled the entire active half of the fire, the only side that had any burning flames. What each crew left behind told something of Park/Forest Service philosophies as well as individual choices by the crew bosses.

The park crew's line stretched for sixty chains, about three-quarters of a mile. Its average width was only a few inches. It was scraped just deep enough to barely stop the creeping fire and it wove its way back and forth through the trees with only an infrequent dead snag sawn and fallen away from the fireline's path. The Hotshot crew left a different trail. They cut only fifty chains, but their line averaged three feet in width, instead of a few inches, and that three feet was scraped clean of every blade of grass or fern. On both sides of this line they cut another four-foot swath, removing every tree under one foot in diameter and every snag even close to their line.

Halfway along their fireline, the Hotshot sawyers had clear-cut a one-acre clearing in case a heli-spot landing site was needed, felling twenty or thirty giant trees hundreds of years old, leaving them lying in scattered rows pointing out from the edge of the potential heli-spot. Their work was excellent, quite professional, but unknown to the crew, a large meadow with a much better landing site lay only a quarter mile ahead, so the big trees died pointlessly.

Where the park crew had been content to let the fire creep up to their fireline, the Hotshots burned out as they went, leaving a secure fireline behind them but devouring large pockets of forest with the intensity of their burnout. Ironically, hours after the total line was completed, heavy rainfall unexpectedly showered down on the fire, extinguishing all but a few smouldering logs and sending both crews scampering for shelter.

If the fire had threatened resources, the Hotshot fireline would far surpass the park crew's for security, but for simple containment of a creeping wilderness fire, the Hotshots left a trail of stumps, scorched forest, and a wide fireline that would be visible for years as man's intrusion into the wilds.

Wilderness controversy really erupts when the firelines get built by bulldozers instead of handcrews. Special administrative permission can allow heavy equipment into wilderness areas to stop major blazes, and especially in the most fragile wilderness zones, such intrusions can leave incredible impacts. A visitor to the Ventana Wilderness behind Big Sur on the California coast will find a complex network of skid trails, firelines, and access roads crisscrossing almost the entire wilderness — all a legacy from the infamous 1977 Marble-Cone conflagration or one of the numerous other major fires that have swept across the tinder-dry brushfields and forest thickets of that wilderness.

When fires can't seem to be caught, it's not hard to see why fire bosses choose four-blade-wide bulldozer lines along the ridges to halt the flames' advance, but such tactics in a wilderness can destroy the very values for which the area was established in the first place. They can leave the image of a cross between a strip mine and a new shopping mall development where natural contours existed for thousands of years previously.

For most wilderness fires, such heavy-handed tactics are neither needed nor allowed, but the tendency for forest or park managers to lean toward intensive suppression over more natural methods has been common. For most firefighters out there on the fireline, there is little concern for wilderness values. If given a choice between tediously chopping and scraping a long fireline through difficult terrain or following along behind a bulldozer, the decision wouldn't even be close. They'd take the big "cat" every time. Fortunately for nature in the wilderness, no one ever asks a Hotshot what he wants.

Like a barren moonscape, the burn becomes an ash-covered wasteland for the crews that cross it.
Author's photo

ten

The Wheeler Gorge Fire

It's already dark when we pull into the town of Ojai, a town many of us know well from many previous fires in the Santa Barbara region of California. Tonight, a big fire's burning a few miles north of town, out in the steep canyons and brush-covered mountains of Wheeler Gorge and the Los Padres National Forest.

Yawning and blinking from the usual lack of any good rest for the past several days, our whole crew expects to be sheltered at some high school or park for the night before jumping to it in the morning. Such expectations are short-lived, however, as sirens wail moments after we enter town and a flurry of fire trucks, police, and frightened citizens speed about in confusion.

After checking in for quick directions, our crew and the El Dorado Hotshots weave through a highway patrol roadblock and head north out of Ojai into a scattering of nice ranches, avocado and orange orchards, and endless ridges of impenetrable brushfields. The fire looks ominous, even from miles away. As we drive toward it we watch spot fires flare up ahead of the fire again and again, like gaseous scouts flung out ahead to find flammable fuels for the main fire. Only minutes after we see the fire, one spot fire ignites almost a half mile ahead of the main conflagration, and within seconds it's spreading both upslope and sidehill with flaming intensity that lights the hills behind it with an eerie glow that foreshadows a miserable night.

Watching the fire, I almost miss the front bus's sharp turn onto a dark side road that snakes past a house, through countless rows of orange trees, and on up to the base of the hill where the spot fire has now turned into acres of flame. Quickly we unload, tool up, and begin scraping line. A tractor arrives right behind us on a big flatbed truck, but even as he begins to unload the fire leaps across the ravine we intended to work and the flames sweep on past toward the town.

Frustrated, we put everything back, reload, and follow the growing convoy of trucks, police cars, and fire trucks back another mile toward Ojai. Right at the roadblock, we take a road to the east, driving past many more orange groves, fancy homes, and dimly seen clusters of frightened citizens. Some are packing cars, others stand dazed, staring bewilderedly up at the nearby hills as the original spot fire spreads with incredible intensity to turn the entire night sky bright gold.

We pull into a long driveway and I'm really speechless. We're at some giant summer camp, and just from my first view I can see more than a hundred people running toward a large central meadow next to the parking lot. Groups of six to ten kids (some still only partially dressed or clad in pajamas) and camp counselors are all stumbling-running, with teenagers crying, camp officials yelling, and everything backlit by the fireworks fantasy skyline display of what looks like the whole world burning up beyond the camp.

Just looking at the fire is enough to know that there's no way of anything surviving that's in its path. Where some big fires move with only a narrow flame front, this monster of a holocaust sweeps across miles of ridges and canyons with a wide flame front that stretches over the farthest hill.

The camp is panic-stricken, with fear magnified even more by everyone being aroused so unexpectedly from their sleep. The Big Guy radios directions after a quick powwow with camp officials and within minutes our crew is running through the elaborate compound, jerking open doors, searching under beds, and looking through closets for any kids that may be hiding from the fire. Everyone's shouting and whooping to make sure no heavy sleeper misses our alarm.

At this point it seems likely we'll lose this camp or at least many of the wooden buildings that hug the nearby brush-covered slopes. Even as we search, small groups of frightened campers still rush past us toward the meadow.

Suddenly a middle-aged man stumbles out of the darkness to frantically grasp my arm. "Billy!" he gasps, "I can't find Billy!" I can see tears streaming down his face and his whole body shaking in violent convulsions.

In seconds I learn a quick description of Billy and the building he's been sleeping in. Ordering the frightened man on to the safety of the meadow, I grab four of the crew and begin tearing through the two-story structure, leaving no bed unchecked or hiding place left unexamined. Like a city where everyone's been mysteriously plucked away, this building and those around it reveal the suddenness of their flight. Clothing's strewn in all directions, bedding's flung back in chaotic disarray, late-night television still drones on unwatched on a portable TV, and spilt cups and broken vases show the effects of still-sleepy stumbling.

We clear every building without finding stragglers, so our crew regroups while I run over and contact the frightened man where everyone's assembling in a giant circle. As expected, Billy is waiting there right where he's supposed to be. All the campers are huddled inside a ring of counselors and staff, holding hands and singing reassuring camp songs while the mountains surrounding the camp on the north and east literally explode with the most aggressive nighttime fire behavior I have ever seen.

Creating its own firestorm, the gigantic fire completely covers every hill visible from camp, each flaming ridge silhouetted against the next higher one until it appears that even the nighttime sky is ablaze. My radio jars my attention back to the camp, still completely untouched, but only for the moment. "Take your bus and start shuttling kids!" calls the Big Guy, so I race back to our crew carrier and pull across the parking lot to join a line of buses, vans, and cars receiving the stream of frantic campers.

Doors jerk open and counselors push in fourteen or fifteen pre-teen and teenage girls, crowding them into the back crew compartment like stuffed sardines. I pull out to the mixed sounds of laughter and crying. It's amazing how different personalities respond to a sudden, traumatic emergency. None of the girls is in the slightest physical danger now that we're leaving the camp, yet three of them are sobbing uncontrollably. A couple of the calmest gently reassure the criers while two other girls laugh senselessly, letting out their own fears in odd, high-pitched giggles. Everyone holds someone else's hand, partly on orders to stay together (where would they go on our bus?) and partly out of an attempt to cling to the physical reassurance of those around them.

We file into the chaos of the nearby downtown high school, unload with a steady stream of fervent "thank you's," and then

I'm headed back to the camp. Already the rest of the crew is spread out along the back of the camp, firing out the hillside in a few spots, but mostly letting the fire work steadily down to the large cleared fuel break that exists along the base of the hill.

Despite the panic and the terror of the campers, not even a single tent or dormitory gets burned. The main advance of the fire sweeps across to the east of the camp, moving on toward the fancy homes and ranches filling the slopes above the main part of town.

Sirens still blare from all the roaded hills beyond us and flashing red lights dot the landscape as the first light of dawn begins to pale the dark background beyond the flames. A slight increase in relative humidity, a slight decrease in the breeze, and suddenly all that's visible for miles around is the thousands of glowing embers, clouds of drifting smoke, and blinking lights . . . the fire's satiated for now.

We leave the camp, drive mindlessly behind yet another convoy of fire trucks and crews until sunrise finds us unloading at the edge of a fancy subdivision, staring in wonder at the erratic patterns left by the now smouldering fire. Charred trailers, with even the tires reduced to nothing but ashes, stand beside houses untouched by the flames.

Iceplant, a succulent juicy ground cover that wouldn't burn with a blowtorch on it, lies almost completely destroyed while patches of bone-dry grasses thirty feet away aren't even scorched. Cars sit gutted far from any other signs of fire. And everywhere, puffs of drifting smoke rise up from thickets of brush and oaks that grow intermingled with the randomly placed homes.

All day we trudge through backyards and empty hillsides, cutting and scraping along the smoking buses and ground litter that still holds burning fire. Rummy from lack of sleep, we hear rumors that the fire's more than 40,000 acres in size, has already consumed seven homes, and somehow has split into three separate fronts—one threatening Ojai, one headed west toward Carpenteria, and one headed north and northwest toward Santa Barbara.

Confusion seems the rule of the day. While the fire lies listless, almost out, no bulldozers are working its edge, no handcrews are cutting line to stop its advance southeast into the outskirts of another part of Ojai. Everyone seems caught up in the mopping-up of last night's damage.

Finally, after too long without rest, our crew's pulled back to firecamp, arriving at Soule Park just in time to watch dry, warm downslope nighttime winds revitalize the fire. Within minutes of the first breeze, a new wall of flames is spreading rapidly across the hills adjoining Ojai, moving eastward and devouring thousands of acres more of grasslands, brushfields, and commercial avocado and orange groves.

Day after day, thousands of firefighters, hundreds of fire engines, and a whole armada of air attack planes struggle to contain this giant Wheeler Gorge fire. Although attributed to arson, there are no arrests made or clear suspects, and millions of taxpayer dollars pour unceasingly into the struggle as the Wheeler fire grows to 70,000 acres. To understand that size, imagine standing on a mountaintop, facing out over foothills and valleys, being able to see a strip of land five miles wide and almost twenty-two miles long. All of that has already burned in this fire.

To the north, only fifty miles away, another massive arson fire scorches more than 30,000 acres. Near Los Gatos in the Santa Cruz Mountains, 15,000 more acres burn with homes lost and lives threatened. Two arson fires burning during this same time period near Los Angeles rip through populated areas killing two people and engulfing thirty more homes. Fire, aided by sick-minded crazies, is on a rampage all over California and many other western states as well.

Eight days after our futile initial attack on this fire, we're driving at daybreak through Ojai, headed north toward the broad northern section of still totally uncontrolled firefront. Even in the faint light of dawn we can read giant "Thank You Firefighters" signs hung up throughout the town. Home after home displays posters or handmade signs with "We love you firemen!" and the stores and banks have tried to outdo each other with praise for our efforts.

Out of all the places and times we've fought fires, this fire in Ojai has brought the greatest public response. Mothers bake cookies in vast quantities for the crews, fruit growers bring in avocados and oranges, children write letters and draw pictures that bring smiles to those of us who, as fathers, know just how much effort and care went into them.

Part of this massive outpouring of affection from the townspeople may have come from the incredible ferocity of the fire and the fact that we somehow have saved all but eleven homes despite many hundreds having narrowly escaped the flames. Another part might come from close interaction, our waiting right in their yards until the flames neared the homes before lighting backfires that rage upslope to meet the main fire and starve it of further fuels. Standing there, beside their homes and landscaped yards, with the air almost supercharged with tension, it's been easy for them to quickly identify with us as well as see firsthand how intensely the bone-dry hillsides burn.

This morning's assignment finds us a few miles north of the narrow, extremely steep canyon where the fire originally started. Along with three other Hotshot crews we pick up where we left off the day before, trying to cut line around one large slop-over where one hot finger of fire stretched down across the highway, whipped out across a half mile of desolate flats, and burned up into a horseshoe of rocky ridges before relenting and submissively smouldering in apparent natural containment.

For many long hours we "cold-trail" the cool edge of the slop-over, cutting line where necessary to reach where the fire crept far under a hillside of brush, but otherwise only hand-feeling along the burnt edge for any sign of warmth. Lunch break finds our crew huddled like a cluster of lizards in the meager shade of one giant boulder, desperately trying to escape the heat and penetrating glare of the bright sun.

I scout out the ridgeline along the cliffs ahead of the crew, hoping to escape from the brutal name-calling and bitter complaints that fly back and forth from one tired firefighter to another. The fire has grown so large that despite more than twenty-five miles of completed, controlled fireline, another twenty or thirty miles of fireline still remain to be built. The thought is anything but welcome to crewmembers already suffering blisters, poison oak, pulled muscles, injured knees, and an assortment of eye injuries.

A half mile ahead of the crew, I reach a good observation point, a small rocky knob higher than the rest of this rocky, knife-edge ridge. Looking far ahead along the ridge I can see that the fire's held up all the way along the rock bluffs, so I radio the good news back to the crew. The Big Guy is jazzed. He wants to get us a burnout assignment anyway, so he sends me with half the crew back toward the main highway two miles away.

I drop off the ridge to meet them, expecting them to eventually catch up to me along the bottom of the lowest ravine. At first the sandy, decomposing bluffs are easy to descend, with long fractures and eroded layers offering good footholds and access. Halfway down the bluffs get steeper and then steeper yet, until I'm really working just to keep from having to climb all the way back to the top to find another route.

I take one long, narrow crack sideways and sloping slightly down across the cliffs, wedge my arms and legs from wall to wall, and descend another forty or fifty feet by pushing and bracing against the rock. Then there's just a long, steeply dropping outcropping, but with my Vibram soles and years of wilderness rock-hopping, I'm not flustered . . . not until I'm most of the way down, that is, and find myself trapped with no cracks to follow, nothing to grip with my hands, and the nearest route a good ten feet laterally across the cliff. I try to turn to go back up, and start to slip. Frozen, I hardly breathe, then too scared not to be reckless, I take two giant sidesteps and leap for the large fissure.

Full of adrenaline, I make it, but stumbling and falling, I miss the edge I'd hoped would halt my leap and stumble out over the end of the outcropping. Fear, no matter what some people claim, is not a pleasant sensation. It only feels good when it stops, and mine doesn't as I half stumble, half fall down a nearly sheer rock face for about twenty feet, slam into a boulder, which I bounce off with a two-handed straight-arm, and then stumble on down in a near-cartwheel, face-near-the-ground frenzy of out-of-control hysteria that ends with a final forceful trip that flings me down onto my hands and knees in a pile of sand.

A few feet either way and I'd have landed on rock and broken bones, but somehow I've half fallen, half stumbled for fifty or sixty feet and only bruised one shin and one hand. My heart still somewhere high up in my throat, I look back up in amazement at the route I plunged down. It's obviously a time to be humbly thankful. Minutes later I look up to see the front half of the crew headed toward me, eager to reach the highway. Rumor

HOTSHOT

has it that a big burn show's about to begin, and our crew's the burn team.

A firefighter trails fire with a drip torch.
Courtesy U.S. Forest Service

eleven

Say Goodbye to the Wildlife

In war, how tough or "macho" a company of soldiers appears depends partly on how deadly their weapons are. In firefighting, big chainsaws or specialized handtools are a minor point of comparison between crews, but Hotshot crews are often judged by other Hotshot crews on how potent their "burn power" is. Our crew, for instance, carries fusees (very similar to highway flares) for day-to-day minor burnout operations when we may need to light an unburned patch of fuels inside the fireline. We also carry napalm canisters and blasters to go with them, small fused explosives that, when screwed onto the cans of napalm and charge, can generate a nice, hot explosion an arm's toss into the fire.

Our specialty, however, is our large firepower of Veri pistols that can shoot magnesium 200 to 300 feet away from the person firing. One burner using a Veri pistol can walk along a road, fire a steady stream of rounds up onto a hillside, and really create some intense fire.

That's exactly our job today, as a whole parade of U.S. Forest Service, California Division of Forestry, and Ventura County fire engines spread out for miles along the highway. The Big Guy and our best burners move into position, then begin firing pistols, throwing napalm, and lighting the road edge with drip torches (metal cans with looped spouts that drip a flammable mixture of diesel oil and gasoline). Within three minutes, the whole slope of the first hillside is engulfed in flames, all surging upslope away from the road and rushing over the first ridge to eventually meet the main fire a half mile or more away.

Such backfire tactics are extremely effective when given an easily held fireline like a highway in an area where endless mountains and continuous brushfields offer little hope of stopping a fire by direct attack. But something about the attitude of the elite

Hotshot crews when it comes to burning doesn't quite jibe with the accepted firefighter image.

Today our crew of burners hurry down the road, competing with another crew's squad of burners to see who can get the most acreage burned in the three to five miles of fireline we hope to establish in this one firing show. Everyone's laughing, yelling, joking, or standing watching with big smiles as whole hillsides flare up in giant fireballs of flame that consume twenty to fifty acres of brush and scattered trees in less than five minutes.

By the time we've ignited the first two miles, the first hillsides have spread their blazes over a couple of ridges and are burning mountains beyond with such intensity that all the new ignitions are sucked in that much quicker by the big fires' drawing in of oxygen for fuel. It's like some jubilant party as hillside after hillside blows up in sudden runs of explosive flames that streak first one way, then another up the mountains. The holding crews all whoop and holler, snap pictures, and generally have a good time—for except for a lot of walking, there's no work for them to do.

This widespread, almost universal pleasure at witnessing an intense firing-out operation reveals a key point about most Hotshot firefighters. Very few particularly care about the thousands and thousands of birds, lizards, squirrels, wood rats, mice, rabbits, and numerous other animals that are being burned to a crisp by their backfire. Often I've watched as firefighters callously chase wood rats or ground squirrels right back into a blazing inferno or laughingly hop about flailing tools onto huddled wood rats or other rodents that aimlessly hesitate on the sanctuary of the fireline.

Beyond a doubt, few Hotshot fire crews show much consideration for wildlife or nature. While a few solitary critics may bemoan the loss of hundreds-of-years-old pines and firs that are being killed by the intentional backfire, most crewmembers are busy cheering the flames on with calls of "Let 'er rip! Yeah! Yeah! All right!" Much like youths given permission to break windows without punishment, these firefighters revel in the destructive power of their torches or pistols, delighting in the charred wastelands left in their wake.

A week later, in the final massive backfire operation that at last, after thirteen days, completes a black-edge fireline around the entire fire, eight more miles of fireline are ignited in one long shift. This great firing operation adds another 15,000 acres to the fire's enormous size, but at this stage fire managers are concerned more about containment than any environmental concerns.

During this final, incredibly hot burn show, we watch hillsides of brush and oaks literally explode with windswept mini-firestorms that tear upslope with such intensity that the brush is burned off at ground level within minutes, leaving nothing but wind-blasted, heat-blistered slopes.

At one point a young doe runs down to the very edge of the fireline, then frightened either intentionally or accidentally by firefighters, she turns and runs back upslope right into the path of the fire. Later, I follow the fresh tracks of a bear that may feel no threat from any fire, since the main fire is more than five miles away from this point along our constructed fireline. Yet minutes later, right where his tracks lead up into a mountainside of oaks, brush, and young firs, I see burners shoot at least twenty pistol rounds and ignite the edge so effectively that not a single bush survives the continuous waves of flaming runs that overlap, crash together, and consume the entire hillside.

For anyone idealistic enough to think that firefighting will save "Bambi" and all the forest creatures, the reality of a big burn show is quite an eye-opener. Such tactics can definitely stop the wildfire and prevent further damage, but somehow that explanation doesn't change the fact that the intentionally set fire kills so many creatures—most uncounted and unmourned—just in one firing operation.

There are the exceptional crewmembers who do make every effort to minimize our damage. Those considered nature lovers may rescue baby rabbits, fat lizards, or bird nests with young to safe habitat, but the few creatures saved are nothing compared to all those lost.

Think for a moment of the consequences of just this one fire, ranging at final count over 110,000 acres, almost 200 square miles of devastation. Think of all the creatures, from butterflies to birds to deer or coyotes, that might live in one square mile—then consider how many more were living in that 110,000 acres before the roaring flames swept over the hills.

Assigned as a burner myself as we near the completion of this final section of intentionally lit fireline, I try to light carefully in an effort to burn an effective buffer near the fireline's edge without creating so much intensity that another wall of flames rips up the mountain. By moving slower and choosing my ignition points carefully, I manage to leave unburned patches up on the slopes as I go, creating safety zones for at least a few animals. But even with care, most of my efforts are futile, for the fuels are so dry and conditions so right for burning that even careful lighting produces one run after another.

As twilight fades and the line nears its final completion, I nearly step on a huge yellow rattlesnake as big as my arm and at least five feet long. Saddened already by the devastation that we've caused, I light beyond him and leave the fire to drive him back toward the shelter of some rocks. Then I flag and sign the area with fluorescent flagging to warn others and move on down the fireline. Just as darkness creates a black backdrop for the flaming burn show, we tie in to the previously burned highway fireline. Two weeks of unbelievable chaos finally end and the Wheeler Gorge fire is left behind for the mop-up crews, forest crews, and county crews who'll spend months more working to reseed with grass, improve lines, and watch for rekindling.

For the Hotshot crews, it's a chance to go home, for however brief a rest, to get clean, make love, sleep, and eat . . . to act like a normal person for a while . . . a very short while. Then it's Hotshots on Parade time again, back to "saving" America's wildlands.

A helicopter skirts the edge of a big "run" of fire.
Courtesy Bob Tribble, U.S. Forest Service

twelve

Our Greatest Fears

At the rate America's wildlands are burning this season, there won't be much left to save next year. Only two days after the Wheeler fire we're flying back into Idaho, unsure whether we'll end up fighting one of the many lightning fires there or go on to ones burning in Montana or Wyoming. Hundreds of small fires are burning all across the upper western states, and three or four big conflagrations have threatened towns, burned homes, and consumed thousands of acres of rangeland.

After seven hours of what seems like the longest bus ride possible from an airport to anyplace in the same state, we arrive at a remote—very remote—firecamp high in the thickly forested mountains overlooking the Salmon River. The actual fire, although surely excitement itself to most of the inexperienced firefighters of that local area, is a typical week-and-a-half affair. First we wait for the 1,000-acre lightning-caused blaze to come up to the river canyon's ridgeline so we can line it there. Predictably, for days the fire only creeps around on the sidehill, until tactics change and crews like ours are sent down into the canyon to cut direct line.

Then the fire somehow gets invigorated, blows out past our pitiful hand lines, and slops on over the ridgeline into a different watershed. For the final four days we team up with a whole herd of Hotshot and forest crews to cut a massive line, burn off from it, and finally contain the flames.

This fire is noteworthy only for its mode of transportation, for the winding dirt road that stretches the ten miles from firecamp to the fire is too rough for the rented school buses that brought us here. Instead we're treated to one of our universal means of moving handcrews: National Guard "deuce-and-a-halfs."

Deuce-and-a-halfs are large trucks left over from World War II (not World War I, despite many rumors to the contrary). The cab holds the National Guard driver and one or two lucky souls who manage to fight their way in or use rank to usurp the padded comfort of a ride inside. The rest of the crew sits in the back of the truck on two long, narrow slatted benches that run along each side of the bed of the truck. Usually seven or eight firefighters can squeeze quite tightly onto each bench, leaving a couple of latecomers the ultimate joy of sitting or lying on the metal truck bed itself.

"Smoothness" and "deuce-and-a-half" are never words that should appear in the same sentence. Once loaded, each crew member grips the bench or side of the truck with a death-grip born from memories of previous rides. Then, with a lurch, the loud, whining, rattling trucks move out in convoy, each driver delighting in staying just close enough to the truck in front of him so the crew in back can taste each billowing cloud of rising dust left behind.

In all seriousness, there are deuce-and-a-half drivers that honestly go out of their way to consider the comfort of the crews, and I can't remember any of them. Instead, most memories are of long, bouncing, bumping rides, whacking tailbones, backbones, and shoulders against the hard benches or slats behind our backs while the crew clowns bellow, "Moooooo! Moooooo!" in long, plaintive calls.

The psychological advantage of such transportation is obvious. Upon arriving at our destination, there is never a time when the crew idly lies back contentedly, reluctant to leave the comfort of our pleasure vehicle. On the contrary, there is usually a mad rush to disembark, often resembling the panic-stricken desertion of a sinking ship as bodies clamber over railings and off the back of the truck in search of solid ground.

One of my clearest memories of deuce-and-a-halfs is of a convoy of fully loaded trucks winding along a steep series of switchbacks during the giant Marble-Cone fire in California. The mountain was so steep and the switchbacks so sharp that the deuce-and-a-halfs couldn't even come close to turning at the sharpest corners. Instead, the driver would aggressively charge up into the switchback turn until the front bumper of the truck almost touched the uphill cutbank of the turn. Then he'd cramp the wheels, slam in the clutch, roll back to the bottom edge of the turn, engage the clutch, and grind on up the mountain.

One turn in particular had duly impressed our crew the first day on the ride in as our driver lurched back on the turn so that the rear bed of our truck stopped just short of a sheer dropoff of hundreds of feet to the canyon bottom below. Being fond of life, we were all apprehensive as we wound up the hillside on the second day, banging and bouncing unmercifully through each series of switchbacks.

Just as we reached the worst turn, our driver accelerated, jerking the truck forward and snapping necks roughly as we all lurched backwards. Our deuce-and-a-half rammed right against the sandy bank on the far side, then the driver cramped the wheels and pushed in the clutch. We rolled back and as he popped the clutch back out, he killed the motor, letting us roll, in a big jerk, back five feet farther before he finally slammed on the brakes.

There we sat, the back half of the truck's bed hanging way out over empty space, the rear wheels only inches from a crumbling road edge and everyone's face turning white with terror. For many long seconds, we hung there as he desperately fought to restart the truck, then with a giant lurch we jerked forward so hard that only the rescuing grip of other hands saved those sitting last on the benches from being thrown out over the abyss.

Such exciting adventures don't overshadow the discomfort of deuce-and-a-half rides, but they certainly improve the stories later. A simple phrase like "We took a ride on a deuce-and-a-half" can evoke a colorful picture of jostling misery to those familiar with such a ride. Thus, deuce-and-a-halfs compare with helicopters for most fear-inspiring ways of getting to the fireline.

Almost every American has personally seen and heard many helicopters, knows what makes them fly, and has seen enough others on television to feel completely knowledgeable about helicopters. Helicopters are a common means of transportation for commuting businessmen, medical emergency shuttles and news media departments, law enforcement agencies and utility companies.

Yet common as helicopters are, few people ever dream of the unbelievable uses that helicopters provide in our new-technology firefighting. And few people would understand why so many firefighters fear helicopter rides far more than the danger of wildfires.

A typical Hotshot fire assignment on a BLM fire in the mountains outside of Las Vegas finds our crew waiting expectantly for a helicopter ride. The dry lightning fire we originally came for burned itself out during the night, but now three or four new fires have started just in the past hour as a number of big thunderhead cells swept across the hot, dry desert, stabbing the sage-covered mesas with indiscriminate flashing jolts.

An S-212 helicopter whines at a deafening crescendo only 100 feet away from where we huddle in two groups, waiting for the hand signal from the heli-tac foreman in charge here at the impromptu heli-spot a mile from firecamp. Finally, all's right, the helmeted pilot gives a thumbs up signal to the heli-tac crewman, and he calls the first load forward toward the open door of the ship. Careful to duck low as they approach, half the crew squeeze in, tighten their seat belts, and then disappear skyward in a quick upthrust of screaming noise that deafens those of us left behind.

Twenty minutes later, it's our turn to lean forward into the blasting grit thrown up by the blades, climb up into the ship, strap in and hold on as the copter whips us up over the desert toward our fire assignment. Helicopters can get a crew anywhere — not just anywhere — anywhere.

In this case we cross ten miles of near-roadless desert in minutes, rise up the sheer cliff sides of a large flat-topped mesa, and land firmly but gently in a sandy clearing only a hundred yards from the burning fire. Admittedly, helicopters save hours or even days of miserable hiking to reach remote hot spots, and even better, they can shuttle water drops from water sources miles away to slow down out-of-control flare-ups or rescue crews cut off from safety.

We're all members of the helicopter fan club on this particular day; the ride out was quick and smooth and they aid us with a couple of nice water drops. Who could ask for more? Six hours later we've completely lined the entire fire and even cooled most of the hot spots still burning inside. Tired but satisfied, we carefully descend the most accessible dropoff along the mesa's edge to reach a large sandy wash a half mile away where we wait patiently again for our ride.

An hour passes slowly, then a strong late-afternoon breeze whips up out of nowhere, gusting erractically but mostly just blowing a little sand around. The 212 flies in minutes later, picks up nine of the crew, and thunders off through the strong winds.

After a longer delay than expected, the 212 finally returns, and somehow all eleven of us squeeze in with the two already on board along with the handtools and all our gear. The screaming whine of the reverberating blades moves up a few pitches, we rock a little, and then we're up, roughly rising in a sharp turn that takes us back to the east. Even with the first violent shudders caused by the wind, we're not too worried until we suddenly twist sideways and find the helicopter slamming ahead all askew and out of control. Frantically the pilot recovers to immediately slam us back the other way, jamming the copter's nose back toward our destination.

We fly like a drunken sailor, lurching from side to side in sudden jerks. With each blast of the wind, the deafening beats of the rotor blades themselves are momentarily swept away by the violent gusts, only to sweep back over us in another staccato wave of sound.

Up ahead, over the pilot's shoulder, I see a high ridge blocking the upper end of the valley we're flying out of. It looms higher than we're flying, and as we attempt to rise, strong downdrafts press us earthward. We climb slowly with the winds easing momentarily, then just as we crest the ridge, blasts of air send us down as if driven by some unseen hand.

Another lurch and the ship climbs reluctantly back skyward, beaten mercilessly by a constant barrage of unexpected gusts that knock us about enough to start producing a nice green tinge in the faces of all the crew looking my direction.

Up ahead another high ridge cuts off the end of the low route we're flying. Our pilot cranks us up on edge in a steep turn to outmaneuver the heady winds that seem intent on pushing us down into the desert. We reach up and over the ridge only to suffer another sledgehammer blast of a strong downdraft that again slams us earthward. Not only do our stomachs rise up to our throats but we watch wide-eyed as the desert rushes up to meet us, then sinks again as the winds pause, allowing the ship to rebound upward in a quick rise that drives our stomachs back down out of our throats and down into our boots, as if on a super rollercoaster ride.

The entire rest of the ride balances between rough air pockets and sudden sidewinds that make everyone think of family, friends, and other ways of making a living. At last we circle, drop low, and land with a rough jerk that we welcome despite our jarred teeth and shaken bodies. The pilot slows, then cuts the motor, gradually letting the blades slow until everything stops. Almost in unison, we all exhale in a giant sigh of relief. Then we spill out into the windy early-evening shadows, thrilled to be alive and grateful for ground to walk on again.

I overhear the pilot comment to his heli-tac crew, "We were lucky to make that one!" and I know exactly how he feels. Crumpled copters are not pretty sights, and those unfortunate passengers in copter crashes usually look even more crumpled . . . not my favorite way to go.

Most helicopter pilots for the Forest Service, Park Service, and BLM are ex-Vietnam veterans who trained and flew countless sorties in the rugged jungles of Southeast Asia. Typically, such sky-jockeys think nothing of hovering right on a narrow ridgetop with only one skid touching the cliff, whining outrageously as they fight to keep the ship level while you and your crew leap out and get clear of the ship.

These same pilots appear equally at ease pounding in through the treetops, their thirty-foot blades narrowly missing giant pines and firs as they slip past the threatening branches to hover over a tiny flare-up, discharge their water drop, and then spiral skyward in a crazy sideways flight that just clears the forest and a rocky ridge above. Confident from years of hazardous flights, such pilots fear little but total lack of visibility from smoke. Yet their very confidence instills dread into unenthusiastic fliers already not too jazzed about helicopter flights.

My own particular concern about helicopters comes not just from wind-buffeted rides, but also from memories of steep canyon shuttles. In steep, narrow canyons, the copters gyrate in tight spirals down and down, not flying flat while they do it, but cocked steeply to the side for maximum control and visibility. It's disconcerting, if not terrifying, to watch the canyon walls spin past in a near-blur of rock, brush, and trees until you suddenly echo past some huge outstretched pine trees to land on a slippery gravel bar right out in the middle of the narrow river. And equally chilling is flying back out once the fire's caught as the ship cranks up, twists sideways, and spins around and around until finally you rise up in a thunderous roar above the canyon rim into the safety of flatter terrain.

Such exciting flying adventures occur all fire season for the heli-tac crew, who unlike many ground-pounders, actually relish the risky flying. Despite the constant chance of crashing in steep terrain, heli-tac crews are always flooded with eager applicants. In fact, on most National Forests, many Hotshot or fire engine crewmen patiently wait for years in hopes of switching over to heli-tac.

Why are copter crews so popular when they face the ever-present dangers of whirling blades, explosive jet fuel, falling from the sky, and countless minor irritants such as the deafening noise of working close to the ships? The main reason is comfort.

While Hotshot crews or forest crews are forever climbing steep slopes laden down with heavy packs and tools, heli-tac crewmen are sitting on the heli-spot high above, watching. While handcrews struggle up through the fading darkness, working with the knowledge that their shift will run all night, heli-tac crewmen are flying back in to the main firecamp (near town) to help shut down the copter for the night.

In fact on many big fires, because pilots and heli-tac crews are responsible for the flying safety of so many others, fire managers make sure that they get comfortable motel rooms and restaurant food . . . quite a contrast to the fate of the regular firefighters who sleep on the ground and eat food cooked by inmates.

Heli-tac crews also love to boast that on fires they really only use one tool—the pencil, for marking flight manifests. Hotshot crews, perhaps out of envy but certainly with some degree of truth, often jokingly claim that a heli-tac crew's only real job is to sit on a heli-spot and drink juices, which is exactly what the Hotshots would like to be doing.

Of course, handcrews like the Hotshots are careful never

to insult their heli-tac friends too viciously. It only takes a minor "error" to get dropped off at heli-spot "C" instead of heli-spot "B" and find yourself with a three-mile cross-country climb just to meet back up with the rest of your crew. And besides, when the old knees or the sore back just can't take another season of Hotshot firefighting, well, maybe there might be an opening for another heli-tac crewman.

Like Hotshot crews, this Indian crew finds a patch of grass is "home" for days or weeks at a time.
Courtesy Bob Tribble, U.S. Forest Service

thirteen

Spice on the Firelines

Picture yourself sitting in total darkness except for the small, narrow beam of your headlamp shining dimly in a circle of light near your feet. Bears, rattlesnakes, and all the other potential dangers of nature don't even cross your mind. Instead, you are acutely conscious that surrounding you in the darkness are forty or so convicted criminals, temporarily out of prison and without any weapon-carrying guard to watch them. Only a few yards downslope you can hear their muffled snickers as one boasts of some prison-yard escapade, while right across the ten-foot-wide fireline from you a 200-pound hulk whispers insults about "pretty-boy Hotshots."

This isn't some bizarre adventure of violence and villains, but a common occurrence on wildfires throughout California and the West. "Con" crews, or inmate crews, consist of state prisoners who've shown good enough behavior and enough enthusiasm to take prison firefighting training in hopes of working off some of their remaining term. And their use on the firelines has become not just a normal event, but a critical source of manpower in battling wildfires throughout the West.

To begin with, despite often sullen appearances or understandably "tough" images, inmates seldom bother other firefighters with any more than the verbal testing common in all street-wise encounters. The same inmate who ridicules Hotshots will usually turn out to be quite friendly and often likable after just a couple of casual replies. I've never even heard of a single instance of inmate-caused aggression at either a firecamp or on the fireline, for the threat of punishment for the inmates is severe indeed.

Most inmates receive only marginal pay for sweating, groveling, and taking orders from their crew leader (usually a Division or Forestry officer) out on the firelines, but the inmates usually work on the basis of one day off their remaining prison term for each day spent on a fire. In addition, they may receive a small amount of pay. But being on an inmate fire crew is "macho," it gives the prisoners a chance to get out from the prison surroundings and atmosphere, and provides them with a lot of freedom compared to those left inside the prison gates.

Most inmates won't bust their butts to catch the fire—they just don't have that kind of incentive. But for mop-up, basic line construction, and holding line, these well-trained prison crews perform needed work at a tremendous savings to taxpayers.

Often while working a fire, we'll work with "mini-con" crews of teenage prisoners who come from the state youth authority camps. As with the adult crews, the majority of the "mini-cons" are there only because it beats the alternative, yet occasionally we'll find a crew fired up with pride and really doing an outstanding job. Perhaps the facing of danger and working together as a crew will reduce these inmates' chances of returning to the prison system, for the ability to get along and cooperate is a necessity every fire crew is forced to learn, no matter who likes whom.

Fighters of wildfire are truly a cross-section of America and its cultures, for it's not just the criminals who appear on the fireline, but many other groups as well. In fact, fighting fire is such an American phenomenon that our country's changing social values can often be reflected in who fights fire.

The original Americans, of course, were the Indians, and appropriately American Indian fire crews have been a longstanding tradition in the West. All-Indian crews, mainly from Indian reservations in Arizona and New Mexico but also from other western states, have been called upon for years to help battle wildfires in the desert region. Also traditionally, whenever major conflagrations have stripped California's firefighting forces, Indian crews have been shuttled from out of state to aid in the suppression efforts.

In the past, many of these traditional Indian crews consisted of mostly "regular crew" type workers—Indians physically fit only for moderate to easy firefighting tasks. But as Indian pride in their culture has rekindled in the past ten to fifteen years, physically top-notch Indians have dominated many of the Hotshot crews from the desert states, while even the "on-call" Indian crews have improved their training, equipment, and discipline to project a high level of professionalism.

Mexican-Americans also have found a cultural tradition as well as high-paying temporary income in firefighting, not just in the desert states but in California's central valley region as well as the more southern sections of the state. On-call "blue-card" handcrews made up entirely of Spanish-speaking crewmen can be found on almost every major Forest Service fire in Southern California during the summer season, often working with a liaison officer to translate and communicate with other crews. They appear at so many firecamps that hot tortillas and spicy salsa have become standard fare at most camp kitchens.

In really hectic fire seasons, besides the Mexican-American and Indian crews, Eskimo firefighters may be dispatched from Alaska to help fill in beside the other crews to fight the flames. Imagine the switch in climatic conditions traveling from 65 degrees and foggy near Anchorage and arriving in Los Angeles in blistering 110-degree heat with intense smog and no shade on the fireline. The Eskimo crews find firefighting especially dangerous when they come, for it's only when blowup conditions have sent firestorms exploding all over the West that manpower lists are stripped enough to justify bringing the Eskimo firefighters all the way down from Alaska.

These varied combinations of ethnic crews can make a large, bustling firecamp seem even more exotic and exciting, with each group's language and culture adding flavor to the already mixed makeup of most regular Forest Service crews. On big fires you can round a bend in the fireline and immediately hear nothing but Georgia twang or Texas drawl as each twenty-man crew from each region yells back and forth. One of the friendliest crews was one we worked with in the Trinity Alps, a crew from the hills of Tennessee that really fit the image of moonshine swigging good ole country boys. But no matter where their origin, each crew works to stop the fire.

To add even further spice to the social mixture of inmates, ethnic crews, and regional differences, the relatively recent addition of women to firefighting again reflects our changing culture.

Yet the transformation of firefighting from a man's world to one offering sexual equality has not come easily, and not without its agony or its humor. Back in 1977, when women were still the exception rather than the rule at fires, one of the two women on my forest crew got so frustrated upon discovering that there were no shower facilities for women that she took matters into her own hands.

The fire we were fighting was the massive Marble-Cone, which burned more than 150,000 acres before it was finally stopped by a force of more than 5,000 firefighters. Our forest crew had been cutting fireline in poison oak and heavy brush for two days, working mostly behind Hotshot crews as we widened their lines. Filthy and already itching, our crew returned to a huge firecamp located along the Hunter-Leggett military base side of the fire, ready for both good food and showers.

Kristin was a college student working for the summer for the Forest Service, and she happened to be quite attractive. Indignant and impatient with the fact that firecamp showers were available for the hundreds of men at firecamp, but none at that time was ready for her or the other few women firefighters, Kristin gathered up her shampoo and soap and headed right down to the showers. Without hesitating she walked right in and showered defiantly in the midst of the men, much to the delight of those showering.

Ten minutes after she'd already returned to our crew's sleeping spot, a large crowd of would-be male showertakers overwhelmed the shower unit, showing how long it takes for any rumor to move completely through a firecamp of 1,000 firefighters or more at any given time. Moments later a flustered camp boss quickly announced over the camp loudspeakers that the showers would be open in the following hour exclusively for the camp's few women.

Today showers for women are a part of almost every firecamp, as is an awareness that women are here to stay on the firelines. But even today it's possible to arrive on a "going" fire and spend days on the firelines without seeing more than a few crews with even one woman.

Part of the answer lies both in the nature of the work and the mentality of those who do it. Take Nell, for example, who's spent six seasons on our Hotshot crew. In her first years, Nell often suffered through firecamps where she was the only woman out of 200 firefighters, half of which were inmates.

First, she had to endure the same physical misery of sleepless nights, gagging smoke, and towering mountainsides that the rest of us faced. But once we returned exhausted to firecamp, she still had to endure the pointed stares of 100 inmates as well as the occasional insults or backhanded slurs from "macho" firefighters irritated by women invading their world.

Nell's particularly even-tempered personality not only helped her to ignore most of the stares and remarks, but combined with her exceptional stamina and good humor, she persisted through season after season while the great majority of our young "macho" firefighters twisted knees, turned ankles, lost endurance, or just mentally couldn't take the strain of the Hotshot life.

But Nell was an exception to the typical female applicant for a Hotshot position. Far fewer women than men even apply, for unlike men, most of whom played firemen as boys, women haven't been culturally funneled toward such jobs. Yet drawn by the same lure of big pay and excitement as the men, the few female applicants who do try out often find the thrill of waking up with their faces in the dirt or going days without bathing not all that exciting. For those women sacrificing their dignity to prove that women can handle the most physically demanding job, it's not the hardships of the actual firefighting but the mental pressure of being so outnumbered by men that eventually sours them on the job.

For those not previously exposed to rough language, the constant banter and crude jokes overheard from their male counterparts are especially difficult to bear, for firefighters and Hotshots in particular have little respect for any subject if they can poke fun.

Of course, we've had such a variety of women fighting fire with us over the years that some of them could out-swear most of the guys. We've had a woman who chewed tobacco and could out-spit the guys who chewed, to a lady on a regular Forest Service crew who wore her hard hat every waking moment for a whole week because her hairdo was messy and she was embarrassed.

But even if they're hard-talking or chew, women firefighters still find resistance. For decades, firefighting was totally a man's world . . . a world beyond the normal social limits expected most places in society. Much like in a rough, lowbrow men's fraternity, many firefighters delighted in a job where they could swear loudly, tell crude jokes, and act without the slightest consideration for the "womenfolk back home." As women in the 1970s began to take advantage of opportunities to join fire crews, these long-standing traditions of rough talk and often outright prejudice conflicted with administrative directives passed down from agency officials.

The Forest Service, aware of some conflicts, began to emphasize the rights of women in employee meetings and newsletters. Some real changes in attitudes were forced onto many reluctant old-time firemen, while others simply adapted by treating women with behind-the-back insults or quiet disdain.

Then in the early 1980s court decisions requiring more employment opportunities in the Forest Service for women coincided with administrative freezes on hiring and advancement. Many male firefighters who'd spent many years on the firelines watched as a few women were selected for jobs to meet sexual parity more in keeping with the country's population. Instead of resenting the administrative freezes that kept almost everyone stranded at the lowest steps on the career ladder, many men developed strong resentment toward the women firefighters rising slowly in their midst.

Much like the civil rights changes during the 1960s, the emergence of women into the male stronghold of firefighting has been a gradual but continual struggle. Yet in a world where danger and exciting drama are often more sought after than wealth, the firefighting field will surely see more and more eager, qualified women applying. And because of the success of those now active, the future should be a little easier for those who will follow in their footsteps.

Sadly, as women do join men on the firelines in increasing numbers, it will only be a matter of time until some women will also die, like their male companions, in a sudden rush of flames.

A sudden flare-up of flames halfway up a canyon wall can mean death to a Hotshot crew caught unprepared.
Courtesy Bob Tribble, U.S. Forest Service

fourteen

Why Firefighters Die

Almost every child learns a fear of fire from touching something hot; fear of a terrifying wildfire should be even greater, and with good reason. Wildfires kill, not just panic-stricken victims trying to flee them, but also firefighters who choose to battle them.

Historically, wildfires have taken a steady toll in our country, with new fatalities reported almost every year no matter how modern our equipment and techniques. Hopefully, the massive death tolls of the past will never be repeated, yet as millions of people move up into the steeper, fire-prone mountains of the West, it seems only a matter of time until a major loss of life occurs again.

Surprisingly, perhaps, the greatest recorded wildfire disaster happened in Pestigo, Wisconsin, in October of 1871. "Cut and run" logging practices had left highly flammable logging slash piled deep throughout much of the state, and tinder-dry weather conditions with strong winds drove massive firestorms over towns and logging camps in overpowering waves of destruction. An estimated 1,500 people died, most of them volunteer firefighters or heroic citizens struggling to save their towns or property.

In the same region in Minnesota in 1918, another series of wildfires erupted with much the same devastating results. More than 500 perished, with again the majority being volunteer firefighters fearlessly fighting the firestorms. The very idea of ill-equipped, inexperienced people attempting to block the advance of a wind-whipped wildfire is almost unthinkable—much like imagining someone trying to stop a runaway locomotive. But fire fatalities haven't always happened at times when strong winds and miles of flame fronts threaten lives. Often they've struck well-trained, experienced fire crews when no obvious danger was apparent.

It was a dry fall in the mountains of the Angeles National Forest in Southern California, but the twenty-one men of the El Cariso Forest Service crew were used to extreme weather conditions. Previously in the season they had fought many wind-driven wildfires under the direction of their high-quality superintendent, who had many years of fighting fires in these same fire-prone mountains.

The fire they were working now had nearly burned itself out during the night in the sparse, scattered fuels on the mountain, so the atmosphere was casual and relaxed despite yet another tedious line-building fire assignment. Far down below their crew along a lower ridge, a county handcrew worked toward them across the bottom of the fire, slowly mopping up hot spots and scraping fireline.

Unlike many other large fires the El Cariso crew had fought that summer, there was no sense of danger from the fire itself, only caution due to the extremely steep slope and slippery footing in the loose rocks on this nearly vertical hillside. They had worked their way carefully down the east flank of the fire, finding a cold or barely smouldering edge as they descended, but still keeping right along the fire's burnt edge so as to be able to move into the already burned section of the slope if any hot spots should start to threaten them.

It was hot, really hot, when the crew finally worked its way on down to the top of a long natural gully or chute that ran up and down the mountain. Within this chute grew only patches of light grass and scattered chamise bushes, with a narrow animal trail made by deer or other wildlife running down the middle.

Leaving the burn edge to take advantage of this easier trail, the El Cariso crew moved into the chute, strung out in a long line as they worked to widen out the natural trail into a four-foot-wide fireline. They had only been working for a few minutes when far below, at a point near the bottom of the chute, a few smouldering clumps of bushes sprang to life, ignited by a sudden strong upslope breeze that began to blow steadily up the side of the mountain.

Because the walls of the chute blocked their view, none of the crew could tell that the hot spot had begun to burn down below. Minutes passed as they continued to chop and scrape in the rocky chute, working carefully to avoid sliding. Then the flames spread into the view of the crew superintendent. Immediately, he yelled a warning to his crew and headed down toward the flames, expecting to be able to knock it down with shovelfuls of dirt. But within thirty seconds, this small fire suddenly flared up tremendously, rushing up toward him so fast that all he could try to do was scramble on his hands and knees up the rocky rim of the chute through the blistering flames.

As he stumbled over the chute edge into safety, his arms severely burned, the exploding wall of flames shot on up the chute, blasting into and over the rest of the scrambling El Cariso crew. The chute's sparse fuels of grass and brush burst into flames, adding even more heat to the fireball that rushed up this natural chimney of a hillside. The entire crew suffered serious burns. Thirteen crewmen died.

Minutes before the explosive run there was no warning, no imminent threat to prepare them for their fate. One minute they were working, joking, and sweating under the glare of the sun. Moments later, the full fury of the fire's run enveloped them in a terrifying wall of flaming death.

The sudden eruption of a seemingly quiet fire has often killed firefighters. The Rattlesnake fire on the Mendocino National Forest in Northern California gives another example of how a seemingly insignificant fire can transform into a consuming fury.

The fire had been burning since midafternoon in brush and grass, its spread aided somewhat by a mild east wind, but quick and aggressive action by Forest Service crews had contained the fire on the west and south flanks, as well as along the top of a main ridge. Besides the already completed firelines, a large drainage and a dirt road seemed to be well located to contain the moderately active fire.

By seven-fifteen in the evening, the fire appeared well under control, except for the occasional small flare-up of isolated hot spot. Under the direction of the fire overhead, firefighters began backfiring from the intended perimeter lines, burning out the unburned fuels between the road and the steadily burning wildfire. About eight o'clock the east winds calmed, leaving conditions nearly ideal for their burning operation. Yet even without the winds, the heavy fuels of thick, old brush mixed with years of dead branches and litter provided dry enough material that a number of small spot fires started from embers drifting down out of the smoke column.

Most of these spot fires were easily "picked up" by the crews holding line during the firing operations, but at eight-fifteen a new spot fire was discovered more than 200 yards down from the ridge in unburned fuels of heavy brush outside the fireline. The slope above this new spot fire was extremely steep and too rough for any tractor to descend to reach the tiny spot fire, so for a while the crews left the fire alone, waiting to see if it would burn on up to the ridge.

Twilight was fading and the fire showed no signs of activity, so by nine o'clock, four firefighters finished with other assignments were sent to start working the tiny fire. It took them almost fifteen minutes just to struggle through the brush down to the fire. Just about the time they finally reached it, an additional force of fifteen more men began hiking downslope to help, also finding their access slow and difficult, but also not taking the time to cut a wide trail into the fire.

The combined forces easily completed lining and securing the spot fire within a short period of time, although without water they could only partially mop it up since it was so hot. While they worked, the fire boss sent five more firefighters on down to help and to carry down bag lunches for the entire force. They took an even longer route through the brush to reach the crews, arriving a little before ten o'clock. By then, with the fire safely contained, the crew leader gladly called a halt and gathered everyone together to break for lunch.

It was very dark and peaceful, with almost no wind to threaten or hint of fire danger. As they ate, the other firefighters over on the main fire began to notice a gradual wind shift, with increasing gusts. Numerous small spot fires began to quickly spring up outside the nearly completed fireline, and one new spot fire started to grow far down near the bottom of the gulch beneath where the crews were eating.

Concerned about its location directly below them, the fire boss quickly ran out on a trail where he could yell across to them to warn them of the fire. At about ten-fifteen he yelled, "Come out! Get out!" until they responded and he knew they'd definitely heard him. Without much enthusiasm, since they'd just started eating, the crews slowly repacked their lunches, picked up their gear, and moved up the slope. Nine of them went up one route while fifteen went a second direction, neither squad finding easy going in the steep brushfield.

Almost two-thirds of the way up, they suddenly could see a glow and hear a roar as the new spot fire below them began a hot run up at their location. Some firefighters threw their tools and panicked, crawling frantically uphill, while others tried to pick the best sidehill route rather than be caught by the first savage runs as the fire rushed up at them. Within a few minutes of their first awareness of the fire, the entire mountainside was overrun by waves of wind-pushed flames. Two of the slowest were the first to fall, caught by the flames as they straggled behind in desperate, out-of-breath fear.

The main group stayed together and nine of them died in a sudden flash-over of fire. Three others perished in scattered locations across the nearly impenetrable brushfield, while one frenzied firefighter managed to flee almost a quarter mile in desperation before finally succumbing to the relentless pursuit of the fire. In all, fifteen firefighters died—casually eating lunch only moments before, then frantically racing the roar, the heat, and the suffocating gases of a suddenly aroused wildfire.

To a casual observer or inexperienced firefighter, almost every major fire might threaten such a deadly ending. To climb up a hillside armed only with a shovel or other tool to confront a five- or ten-acre fire with its sudden flashy runs, noisy bursts of flame, or dense blankets of obscuring smoke may seem nearly suicidal. Yet it's really quite safe . . . if you know fire behavior. For today's firefighter, the danger is still possible, but far less likely, since agencies like the Forest Service provide the safest, most modern personal protective equipment and technology available.

Tragic wildfire fatalities like those mentioned above have produced some amazing changes in both gear and communications. Fireshirts and firepants are made today from Nomex material, a specially treated lightweight fabric designed to resist the hottest flames. Although every Hotshot will agree that such clothing still doesn't stop the heat from penetrating, these uniforms won't ignite and take direct flames with only minor scorching.

HOTSHOT

Modern hard hats, safety glasses and goggles, various types of gloves, and high-topped, Vibram-soled boots round out the basics of fire clothing. Boots are a special mark of an experienced firefighter, with one or two expansive but top-quality name-brand styles chosen by the overwhelming majority of firefighters. Like artists evaluating a work of art, Hotshot crews will often cluster around a newly arrived pair of $250 boots to praise, criticize, or compare to their own. Besides contributing to their image, high-quality boots can make a great difference to a Hotshot's feet after hundreds of hours of climbing mountains, hiking ridges, or standing in hot ashes. Woe to the luckless firefighter whose feet hurt too badly to finish a fire shift. He may be flown out by copter or allowed to evacuate, but the years of painful insults he'll suffer will far outlast the misery his blisters may have caused.

Besides modern clothing, the one real life-saving tool worn by all Forest Service firefighters is their fire tent or fire shelter — a thick aluminum-foil/paper tent that's required on all active fire shifts. Known as the "baked potato bag" or "shake and bake" or various other insulting names, this lightweight shelter is usually worn on the web gear or waist. These shiny tents can be opened and draped over a firefighter in less than thirty seconds, catching a tentful of breathable air as the firefighter drops to the ground to anchor the pup-tent-shaped shelter over him with his feet and hands.

Light straps help the firefighter hold the tent down when violent fire winds blast against it, and its reflecting surface may keep the interior as low as 200 degrees when the air outside is more than 1,000 degrees. The air trapped inside the tent is also vital, for often it's the superheated gases of the firestorm that scorch the victim's lungs long before any burns have proven critical.

Lying face down inside one of these shelters using worn-out tents in a training session, it's easy to imagine the terror of actually using one. Hiding in the dark interior, the firefighter can't see the approaching wall of flames, he can only hear its noise, feel its winds on the tent, and eventually feel the tremendous heat roasting him inside like a Christmas turkey. But experience shows that these tents definitely work — *if* the crew has time to pull them, *if* they have a large enough clearing to actually deploy them, and *if* the fuels amongst them are minimal. Using a tent in a solid brushfield or amid thick piles of heavy forest slash would only postpone the inevitable outcome.

But the encouraging part of equipment like these fire shelters may be the fact that seldom do firefighters ever need to pull them.

Today, most Hotshot crews carry four to six multi-channel portable radios spread our among their twenty members for maximum communications and fireline safety. Forest Service safety standards have risen dramatically in the past two decades, with mandatory week-long wildland fire training required for even temporary firefighters. Additional, in-depth training courses move up in both range and complexity, with crew leaders today required to be able to use computers, field guides, and on-site instruments to accurately predict expected "rate-of-spread" or fire size at various stages of a given fire's expansion.

Overhead in the nighttime skies, special planes fly over big fires with infrared cameras, using their heat-sensing ability to pinpoint exactly on aerial maps the location of hot spots along the fire perimeter. On the ground, small night scopes also feature heat sensors to visually detect heat as insignificant as a lit cigarette, while ultra-modern tools the size and shape of a flashlight beep offensively whenever they're pointed at underground pockets of smouldering roots or glowing embers.

Combining all these state-of-the-art instruments with high-quality training can only better prepare fire crews for dealing with the still-often-unpredictable "demon" fire. Yet surprisingly, despite all of modern man's advances and techniques, the battle to halt a raging wildfire often comes down to a very primitive struggle of frenzied men using simple handtools to fling dirt and slash brush as they fight to smother and starve the flames. And sometimes, in that battle, the firefighters will lose . . . not just a mountainside, but also their lives.

A burner leaves a trail of diesel-gasoline mixture burning behind him.
Author's photo

fifteen

The Good Side of Fire

Dark-gray clouds cover the sky as our crew pulls back into our warehouse parking lot after yet another off-forest assignment. Shivering in the first cold spell of fall, we bustle about unloading and refurbishing our trucks, hurrying to finish before the rains come.

Despite their foreboding canopy, the clouds drop nothing until late that night, when triggered by just the right natural conditions, they suddenly unleash a heavy downpour that lasts for more than six hours. The intense storm soaks our central Sierra Nevada region with almost two inches of moisture, filling the streams, wetting the soils, and watering the thirsty forest vegetation that's waited four months since the last rains. Forest officials declare local fire season officially over; state fire crews start planning staff reductions, and our Hotshot crew begins lighting the forest on fire.

Lighting the forest on fire, when twenty-four hours earlier we were fighting to extinguish every flame, is not the contradiction it first appears to be. A Hotshot crew, along with other Forest Service fire crews, spends a good percentage of its season lighting fires rather than stopping them, for fire is *good* for the forest.

Despite decades of Smokey the Bear admonishments to prevent forest fires, forest researchers discovered about ten to fifteen years ago that the forest environment actually depends upon fire to maintain a healthy, balanced ecosystem. The fire that is necessary, however, is not the wind-driven, all-consuming wildfire of summer, but a well-controlled, low-intensity fire during mild to wet weather.

Prior to modern man's mastery over the forests, natural lightning fires burned periodically throughout most forest regions, consuming much of the dead branches, leaf litter, and debris of fallen trees or standing snags that built up between fires. These frequent fires also burned up most of the brush and young trees that sprouted every year in the forest, leaving the biggest, most mature trees only scorched while creating a natural mosaic of scattered clumps of bushes and young trees. This frequent natural burning created, over thousands of years, a majestic forest of giant towering conifers and hardwoods standing amidst patches of small trees, scattered bushes, and dense ground covers of quick-spreading fire-tolerant ferns, grasses, and wildflowers.

This was the forest of the 1800s when John Muir and other explorers could ride at a gallop for miles throughout the mountain forests. But well-intended intervention by man, especially in the period from 1920 to 1980, suppressed wildfires and completely halted the natural consumption of fuels and the "opening up" of the low and middle levels of forest fuels.

The result was sixty years or more of a tremendous buildup of forest debris and thickets of young trees and brush. Not only did such conditions create catastrophic fuelbeds for uncontrollable wildfire, but this one-foot to four-foot layer of dead branches shaded out or buried the succulent ground covers so necessary for forest wildlife.

In the 1970s when forest managers began to learn about the need to allow fire into tree stands, they also discovered new problems. Budgets were tight, and while administrators willingly paid to stop raging wildfires, it was much more difficult to find funding to light "good" fires. Another problem was teaching the public, especially local communities, why fire was necessary, rather than always bad. And the final problem dealt with how to use fire to burn just enough and not too much.

Accordingly, "prescription" burning was officially born, a method of using experience, calculations, and measurements of local fuels and weather to determine exactly when to burn and when to stop. In the national plarks, such as Yosemite, administrators also began to allow natural fires to burn unchecked when they were located above certain elevation levels or away from campgrounds, trails, or developed sites. By monitoring these natural, slow-moving fires, naturalists and forest managers found these burns revitalizing important wildlife habitat and actually increasing the scenic quality of many high-mountain areas.

The Good Side of Fire

Within commercial timberlands, fires began to serve two distinct purposes—one for timber production and the other for wildlife. Our Hotshot crew works on both types each year, once conditions moisten the flashy fuels like grass or pine needles enough to prevent out-of-control flare-ups.

Wildlife fires are often funded by state fish and game agencies to treat hillsides of overgrown brush in areas where deer concentrate. We light these fires with just enough intensity to open up the brushfields without killing most of the oaks or conifers that also grow in scattered clumps across the slopes. Using previously constructed, wide, safe firelines we can burn hundreds of acres and create much the same mosaic pattern that natural fires often make.

Burning the brush also causes heavy production of new sprouts much higher in protein than the older leaves. These tasty new sprouts not only appeal more to deer, but their higher protein levels may make the critical difference in whether or not they survive the rigors of the strenuous winter season.

While wildlife burns are designed to mimic nature, our timber management fires work to wipe out the natural system present on a site and open it to timber production. For a Hotshot crew, these intense burns may be more work than many wildfires.

"Bad news," sighs Milo, greeting me as I store my lunch on our crew bus prior to work. "We're burning the steep clearcuts today."

For Milo and the rest of our bus, that is bad news, for we'll be assigned to holding the smoky, hot upper fireline while the other half of our crew lights the logging slash down below. This could be a long project day.

Once out on the project site, Nell, Milo, Dave, and I take the upper fireline and spread the rest of our holding force down along each flank of this sixteen-acre clearcut unit. Like much of private industry, the Forest Service has turned to clearcutting as its preferred method of logging on the national forest, despite often intense opposition from environmental groups and local citizens. The Forest Service sells cutting rights to lumber companies to strip all the big trees from sites like this one, then either logger crews or Forest Service employees cut down all the remaining young trees, oaks, and brush, letting them dry in place.

When these fuels of waste wood and slash are cured, crews like ours burn the entire clearcut, trying to burn it hot enough to consume not just the logging debris and waste wood, but also young brush and ground covers that might later compete with the crop trees. Once the site's bare of competition, rows of pines or firs are planted with uniform spacing across the site.

This morning's burning goes smoothly, with only slight breezes scattering occasional embers as the fire's column rises straight up into the sky. Then Nell calls out a warning, "Looks like we're getting a wind shift!" and sure enough, the smoke lies over against our slope, immersing us in its depths.

Dave stumbles out first, pleading inability to breathe, and Nell and Milo are soon behind him, half feeling their way along the logging road that separates the clearcut from the healthy forest just on the other side. Taking turns, we cross in quick runs along the road every few minutes, scooping up embers or smouldering hot spots with our shovels and flinging them back into the burn.

Breathing smoke isn't much fun to start with, and it's even less fun when our crew's the one making it. During one particularly bad spell when none of our holding squad can do much but gag, I whine and complain over the radio, but all to no avail. Someone has to hold the line—today's just our day.

The same complaints about smoke can often be heard by local homeowners, visiting campers, or tourist promoters upset over brown skies and stinking air. When up to 400 such clearcuts may be burned each fall on just one national forest, smoke levels can often reach unpleasant or annoying levels in the foothill communities that border the forests.

Yosemite National Park's aggressive efforts to restore the natural habitat of its pristine parklands by the use of fire have proven quite successful, but occasionally down-canyon winds have pulled smoke downslope into the populated foothill zone and provoked criticism, especially from the elderly. Most years, however, Yosemite's burn program has functioned smoothly, with all the smoke pushed higher into the mountains or dissipated by the dominant west winds.

Yosemite is a prime example of how the use of both natural and "prescribed" fire can improve or restore habitat. Steve Botti, resource management specialist for the park, oversees a controlled burn program that burns an average of 2,500 acres every year.

"I feel very strongly that prescribed fire justifies its costs, even if you just look at its benefits in reducing wildfire suppression. We spend about $15 an acre in our controlled burn program, while it costs the Park Service somewhere in the neighborhood of $2,000 an acre to suppress wildfires.

"Our main emphasis with controlled burns is to reduce the hazard fuel loadings, to make it more like it once used to be. Our program will gradually recreate the open parkland and large-tree forest that the early explorers describe as being so wondrous," explains Botti. "Esthetically, it should be much more pleasing and it'll provide a much better balance of habitat for wildlife."

Our fire today on this clearcut unit burns much hotter than the low-intensity fires of Yosemite, for one to three feet of logging slash covers much of the site. Even as the smoke begins to clear and we begin checking the perimeter prior to leaving for the night, we know that this unit will smolder for days and weeks, continuing to creep around within its well-placed boundaries.

Hopefully, no extreme winds or hot spell next week or next month will unexpectedly turn this and our many other burn units into a threat to other forest resources. Such unusual quirks of weather do occasionally occur, sending fire crews rushing in all directions to find out which smoke is inside the burn lines and which is outside.

Part of the irony of working in fire is the sudden switch from lighting them to fighting them. Almost every year we're called away from mild-weather burning projects in our snowcapped mountains to return to Southern California to battle wind-driven conflagrations. These dangerous fires, often pushed by Santa Ana winds of fifty to more than eighty miles per hour, prove to be classic fires in both magnitude and memories. Fires as large as 40,000 acres may rip through foothill or mountain communities at the same time that heavy snowfall or relentless rain in the Pacific Northwest may be flooding communities there.

These fall wildfires are also memorable in that late season layoffs or return to college by some crewmen leaves Hotshot crews like ours shorthanded. When the big fires break, we fill out our empty slots with "old-timers," ex-Hotshots who've moved into management jobs, fire prevention work, or even into helitac positions. These old-timers eagerly head south, glad to make some big overtime pay without having had to endure our long season of fire or put in the long hours our crew spends on project work. And like Hotshots on parade, classic Hotshot crews from each national forest will file into firecamp, filling the spare moments between fire shifts with lots of socializing, catching up on life's changes, and nursing aching knees and backs.

Then crowded around the few outdoor heaters in the cool fall air of firecamp, these Hotshots will laugh and shout and then grow serious as one firefighter begins to weave a tale of some past fire exploit. And everyone will listen, lost in their own memories of bursting lungs and burning heat and fear. They'll listen while someone tells of a fire like the Indian fire.

Against a backdrop of flames, firefighters battle the fire's advancing edge.
Courtesy U.S. Forest Service

sixteen

The Indian Fire

The noise of my chainsaw drowns out whatever Milo is trying to tell me, so I switch it off and pull out my earplugs, glad to ease the strain on my ears. In the sudden silence Milo gestures excitedly, "Look at that!" pointing far to the northeast from our work site here.

At first all I can see are a few large cumulus clouds that tower high over the sharp backbone of the Sierra Nevada range, then I realize that there's something odd about the biggest cloud.

"That's incredible!" I agree, amazed that a fire has grown so quickly and with such force that it resembles a 10,000- or 20,000-foot-high cloud. Only the darker brown color of the smoke column gives away its origin—that and the sheared-off top, boiling white in freezing contrast to the rest of its color.

Without even waiting for orders, we pass the word across the hilltop down to where most of the crew's working piling cut brush and scraping future firelines. "Everybody back to the trucks! They've got a real gobbler over on the east side!"

Not even five minutes after we reach our trucks, the first report comes over our radio. "BLM's broken a big fire south of Carson City . . . with winds gusting up to forty miles per hour and no air tankers available at this time." We stuff our gear into the trucks and head in toward the highway, nearly reaching town before our call finally comes.

The Big Guy is already waiting for us at the highway, eager to get a jump on the two-hour drive over the steep, winding mountain pass and on up along the east side of the range to the fire. It's a beautiful drive during the summer, even if a raging wildfire waits at the other end, and it's a treat to make the trip in the daytime instead of in the dead of night.

By five o'clock we're zipping at top speed for our crew buses along the base of the sheer east slope of the mountains, headed north on highway 395. Minutes later we get our first real view of the fire, just a quick glimpse of black smoke and a red glow eating its way down out of the foothills toward the flatter desert ahead of us.

A few miles south of the small town of Gardnerville, the highway runs down between high rolling hills covered with a thick forest of pinyon pine and junipers, then straightens out to stretch for miles across a huge desert plain of sagebrush and grasses. Far to the west, on the very edge of this broad plain, the rolling dark clouds of the approaching fire obscure the entire sky. It's like a movie setting for Dorothy in the *Wizard of Oz* just as the tornado's about to strike.

Monitoring radio traffic as we drive, we're aware of how confused and apparently undermanned the fire agencies in this district are at this time. Other fires have broken out in previous days to the north of here toward Reno, so many crews are already committed to those blazes. Our own official destination is Carson City, but after we drop down on the highway into the broad plain, we come over a small rise just south of Gardnerville to find a whole armada of fire trucks, highway patrol cars, sheriff's vehicles, and curious sightseers lining the highway for a quarter mile on both sides of the road.

Stopping to tie in with a single BLM unit, our crew quickly appreciates the breathtaking view that's frozen so many people to this stretch of road. The entire fire is finally visible to the west, reaching from north to south along the high foothills and across the plain in a spectacular, otherworldly landscape. The dark smoke cloud, miles wide and bent over sharply toward us, leans far out over our heads, miles closer than the slower-moving flames that may only be spreading at twenty or thirty miles per hour, rather than the great speed of the dark clouds rushing past high overhead. The sun is setting in the west behind the high mountain range, and its colorful descent backlights the smoke at the same time that firestorm conditions along the flame front flare out in equally brilliant reds and oranges.

The advancing flame front is still many miles from our position along the highway, but none of the county, city, or volunteer fire crews waiting nervously appear to be preparing for any kind of offensive action. Perhaps the fact that the fire looks to be at least two miles wide along its flame front is so overwhelming that no one believes that any human action can even affect such a massive wall of flames.

The Big Guy talks quickly with one BLM boss and a few other overhead types from the other agencies. Yet no one seems to want to be in charge. The threat is just too overpowering. We can feel the strong, hot wind rushing over us. We can already smell the smoke even though the fire's still miles away, and looking north toward the little town, it's hard to imagine any fate but disaster. We won't have to go out to fight this fire. It'll be here soon enough.

Almost an hour after our arrival, we suddenly get the word to load back into our buses and move up the road to the south end of town. The highway runs right through the town, dividing a few houses, ranches, and a trailer park from the main cluster of homes on the east side of the highway.

Paul calls me over to the Big Guy's truck and quickly diagrams for Frank and me what we'll try to do in the few minutes before the fire sweeps over the residences on the west side of the road and hits the highway. Pointing out the river running along the west edge of the inhabited area, he shows us where we'll try to build a fireline to backfire off from to try to save the homes west of the road.

After the long wait, we're all anxious to get going and everyone throws himself into the work with adrenaline pumping. We can actually feel, as well as hear, the fire getting closer, now only a mile or so to the west, orange flames licking upwards all along the base of the black cloud, almost lost in the darkening night. We can also see people scurrying about at the threatened homes, throwing clothes and valuables into pickup trucks, while the first few fire engines begin to move into the area, ready to do what they can when the flames arrive.

More fire engines, with sirens wailing, drive up every few minutes as volunteer and city rigs from all the nearby communities rush to aid this threatened town. Unlike most of our fires, which threaten only natural resources and occasional homes, this fire looks like a major tragedy in the making. The tension's heavy, almost desperate, as we beat our tools into the rocky, sandy dirt, chopping the thick, tough bone-dry sagebrush at ground level to open enough of a fireline to light off from safely.

We're only three-quarters of the way to the river when the Big Guy orders our burners to begin lighting from the highway along our line. The fuels are incredibly dry, with flames shooting ten feet into the air off bushes only three to four feet tall, but the blowing winds bend the flames back across our line, singeing bushes on the other side and scattering hot embers everywhere.

Hotshots run about in all directions, slamming shovels down on igniting grass patches, flinging sand, and chopping at burning bushes. I move up with some of my squad to finish cutting the line, aware that even one weak spot and the fire will breach our puny defense and race through the sage-covered slopes where the homes sit. The winds whip our backfire against our line and suddenly I hear shouts. Looking back toward the road, I see a spot fire blaze up brightly. Frank yells for one of the small county trucks to pull down and aid us with some sprayed water, but afraid to drive into bumper-high sagebrush in case it ignites, the fireman stubbornly refuses.

Jack and Nell take two other crewmen and attack the flare-up while even now the burners continue to light more brush along the line, trying to beat the threatening main fire. Just as we tie a lousy scratch line down to the river, Kurt and another burner light off the last stretch of brush, sending a whole hillside of flame leaping directly toward the onrushing main fire.

The backfire seems overwhelmed as the leading edge of the far-bigger fire surges against our intentionally lit wall of flames. Joining together, the two fires push flames thirty or forty feet into the air, then the thundering main fire, driven by the wind, pushes on toward the road, barely deflected from our fireline by our just-in-time backfire.

Gasping for breath and still not sure we've held this one small flank of the fire, we move up to a high point along our line to watch the main fire. All the other fire agencies are supposed to hold or backfire off from the main highway, but lacking firing equipment and backfiring know-how, they wait too long. Even as we watch, the main fire hits the highway and roars across from cutbank to cutbank one hundred feet horizontally through the air in a fiery leap that outflanks ten or twenty rushing engine crews frantically trying to wet down countless spot fires all across the east side of the highway.

"On the double!" yells the Big Guy, and we run as fast as we can in the smoke and bad light of our headlamps and the fire. Back to the highway, we line out again, cross the road, and begin lighting off from someone's dirt driveway leading to their house. A couple of huge bulldozers suddenly whine out of the chaos, scraping a double-blade "catline" directly east across from our successful fireline on the other side of the highway. If we can hold the fire on their catline, we can keep it out of the houses, barns, and outbuildings that fill this entire neighborhood.

Terror is on the wind as well as fire. Sirens are wailing now in every direction as engine after engine rushes into the subdivision to begin protecting structures. Horses are screaming in fright, galloping wildly in fear around their sage-filled pastures while their owners or firemen futilely attempt to calm them. A lady's shouting from her porch; what, I can't say, for we're overwhelmed by the howling wind and the roars of speeding engines and the scraping tractors.

Kurt works with the bulldozers, getting directions on his radio from Paul or the Big Guy who lead us to the first threatened houses. Pointing to the spot he wants us to start, the Big Guy yells, "Cut line and burn off from it!" before moving into the smoke to scout ahead.

The wind seems to pick up even more intensity and shifts more toward us, driving the now roaring fire on this side of the highway straight at our threatened homes. Without even slowing, the flames sweep across sections of the catline, but our new backfires suck in toward the main fire and consume most of the fuels before the main fire can reach them.

We chop and fling sand at flaring hot spots, lighting sage right next to someone's beautiful wood-sided house, throwing more dirt as burning leaves and bits of dead wood shower against the luxury home. Then Kurt brings the tractors right up to the houses, turning them and directing them as they carve a slashing fuelbreak across the front of another eight or ten homes. Even with the scream of the sirens driving us in a frenzy, we can't keep up our hot-spotting pace much longer.

I watch as one of my squad loses his grip on his tool and

narrowly misses cutting his foot as it slips and bounces up off the ground. Yelling at him, I'm cut short as I look past our frantic crew to see a woman, standing silhouetted in the dim light of her window, watching the fire with one arm wrapped protectively around a child. Why she's crazy enough to still be here is beyond me, but her rigid fear only underscores how critical our efforts are right now.

I'm yelling encouragement, Frank's leading the saw team and burners at a fast pace, and Paul's rushing back and forth, sending one or two crewmen off to pick up spot fires every few minutes.

Then the bulldozers are right amongst us, turning this way, then that, so quickly in the wind-driven smoke that it seems certain someone's going to be crushed. Kurt leads one tractor off to the east, hooking ahead of the fire as the winds ease just enough for fire engine crews and volunteers to catch the many spot fires threatening houses even a block or more in from our fireline. So far we're holding the flames—no homes are more than slightly damaged and our burners have reached the turn in the catline. We might be able to divert it from the town yet.

Satisfied we've got this one area secure, I send everyone but Nell and myself up to work the backfire leading to the east. An old pile of tires catches fire, and after yelling myself hoarse, I finally get one of the big city fire engines to move up and douse the pile with hundreds of gallons of water. I look back at Nell's cry and there, across a deep gully, a spot fire's spreading toward another section of homes. The rest of our crew's already far ahead, almost running to keep up with the burners and the tractors. I radio for help and while Nell and I struggle to hold the blisteringly hot head of the fire, try to plan another spot up ahead to fall back to if our efforts fail.

But Milo and a squad of four arrive like the rescuing cavalry, running back through the darkness to join in the attack. It takes us thirty minutes or more to knock down the flames, cool the hot spots, and build a secure fireline, but by then the winds have dropped to "mild" breezes of ten miles per hour or less. Sagging but feeling really proud of our crew, I radio that the spot fire's contained and that we're hurrying to catch up. The backfire seems to be working well across a wide flat, with the catline hooking the fire's head and turning across it up along a long ridge that runs south, parallel to the highway but two miles east.

The Big Guy and Paul send two men back for all the rest of our burning gear, taking advantage of the dwindling winds and the slowing advance of the fire to give us a ten-minute break. Kurt and the tractors move out ahead, bulldozing a wide fireline for us to burn off of along the juniper- and sage-covered ridge. The burners light quickly, creating as much fire downslope from the line as possible in order to pull the burning edge down from the fireline rather than let the west wind blow the fire across the line.

All night long we struggle to hold our burn show, with our back half of the crew always suffering from dense, choking, gagging smoke as we stand in the hottest sections along the fireline to pick up the constant spot fires that light so easily in the bone-dry desert fuels. Over and over I drop to my knees, searching for the barely breathable air that may still exist flush against the ground. By dawn we're stumbling, not just from lack of sleep but even more from carbon monoxide inhalation from breathing far too much smoke.

As sunrise's first glimmer of red lightens the morning sky, the tractors turn west toward the highway, completing the hook as we burn back toward the road. Two hours later we totter, gasp, and sag to a stop near the highway's edge, exhausted, but successful in stopping, then lining this seemingly unstoppable conflagration.

No one's applauding, no one's cheering, least of all our crew. We're too spent, too sick from the smoke to even want to sit up, but in another way we feel great. We know that our Hotshot crew, composed of twenty individuals, has worked together to face an overpowering natural force . . . and we've won.

No matter that our backs ache or our lungs are scorched. We've taken our skills and training and saved people's homes, saved their pets, and maybe even a few lives. We've also tested ourselves once again and felt good about how we measured up.

I squint out of stinging, swollen eyes at those around me, most of them crumpled in semiconscious contentment at this moment of rest. Then Nell and Milo both smile back at me, shaking their heads as if to say, "Why do we do this job?" But

HOTSHOT

I know the answer already. We do it because we're the Hotshots.

John Buckley spent thirteen years fighting forest fires for the U.S. Forest Service, six of those fire seasons on Hotshot crews. He now lives in California with his daughter Lara and son David, working with local conservation groups to protect threatened natural areas. He also works with homeowners, schoolchildren, and forest visitors to prevent wildfires.